Kristen Nehemiah Horst (Ed.)

Maldon Railway Station

Kristen Nehemiah Horst (Ed.)

Maldon Railway Station

Victorian Goldfields Railway, Bendigo railway line, Castlemaine, Victoria, Victorian Goldfields Railway

Dign Press

Contents

Maldon railway station

Station Information	
Line	VGR former Maldon Line
Distance from Flinders Street	143.4km
Number of Platforms	1
Number of Tracks	1
Station status	Tourist Station
Station opened	16 June 1884
Station closed	3 December 1976

Maldon is a historic railway station on the Victorian Goldfields Railways Maldon branch line, off the main Bendigo, Echuca and Swan Hill lines in central Victoria, Australia.[1]

History

The station was originally opened on 16 June 1884 and was closed to passenger services on 6 January 1941.[2] After this date, the line was used only for goods traffic until closure on 3 December 1976.[3] The Maldon station was opened for tourist services in 1986 over a short 1 km section of the line out of Maldon. The line has been extended to Castlemaine.

On 20 October 2009 the roof, kitchen and stationmaster's office at the Maldon Train Station were extensively damaged by fire.[1] [4]

References

[1] "Historic station damaged by fire" (http://www.news.com.au/story/0,27574,26240836-1702,00.html). AAP. 21 October 2009. . Retrieved 21 October 2009.

[2] "VGR Timechart 1941-1960" (http://www.vgr.com.au/history.html). www.vgr.com.au. 2005-02-27. . Retrieved 2007-01-17.

[3] "History & Preservation" (http://www.vgr.com.au/history.html). www.vgr.com.au. 2006-09-16. . Retrieved 2007-01-17.

[4] "Maldon railway station damaged in blaze" (http://www.abc.net.au/local/photos/2009/10/21/2720151.htm). ABC Local News - Central Victoria. . Retrieved 21 October 2009.

Station Navigation	
Victorian Goldfields Railway	
Muckleford	Previous Station
Entire line	

Victorian Goldfields Railway

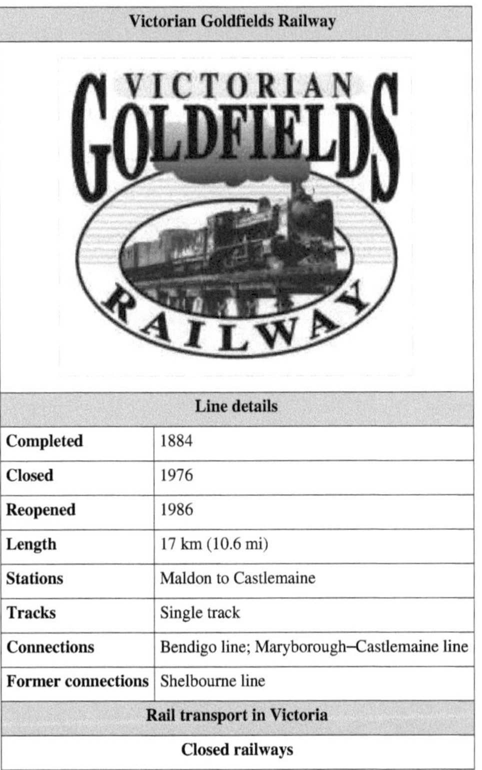

Victorian Goldfields Railway	
Line details	
Completed	1884
Closed	1976
Reopened	1986
Length	17 km (10.6 mi)
Stations	Maldon to Castlemaine
Tracks	Single track
Connections	Bendigo line; Maryborough–Castlemaine line
Former connections	Shelbourne line
Rail transport in Victoria	
Closed railways	

The **Victorian Goldfields Railway** is a 1600 mm (5 ft 3 in) broad gauge tourist railway in Victoria, Australia. It operates along a formerly disused branch line between the towns of Maldon and Castlemaine.

J 515 at Maldon, awaiting departure for Castlemaine, 20 January 2007

History

The original line was opened on 16 June 1884,[1] opening up rail access from the established station at Castlemaine to the towns of Muckleford and Maldon. The area was prosperous, as Castlemaine and Maldon had both experienced gold rushes in the preceding years, and local residents had been petitioning the state government for a railway since 1874. On 2 August 1884, a contract was let for an extension to Laanecoorie, however further construction was suspended after the line reached the small town of Shelbourne in 1891.[2]

The line was served by twice-daily trains for the first forty years of its life, which was increased to four-times-daily trains in 1924.[3] However, these were cut back at the end of the 1920s due to a decrease in the local population, and passenger services were eliminated altogether during World War II.[4] This meant that the line was only used by a weekly goods train which went through to Shelbourne. When bushfire damage caused the closure of the Shelbourne extension in 1970, the remainder of the line was rendered largely useless, and it was officially closed in December 1976.[5]

The response to the closure from the local community was swift, and the Castlemaine and Maldon Railway Preservation Society was founded in the same month, with the intention of reopening the line as a tourist railway. While Maldon station was intact, and was able to used as a base for their operations, they were faced with numerous problems: a line that needed substantial repairs, a lack of rolling stock, and rebuilding the demolished station at Muckleford.

Reconstruction

Over the next decade, volunteers obtained and renovated rolling stock and by 1986, trains were able to operate on a one kilometre section of track out of Maldon.[5] By 1996, the line from Muckleford to Maldon had been restored, with the platforms being reinstated and a small replacement building being constructed.[5] Services were able to operate along a regular timetable, and the society set about reopening the Muckleford-Castlemaine section of the line.

First train from Maldon approaches Castlemaine station, 19 December 2004

The project received the support of the local,[6] state[7] and federal[8] governments. The necessary physical work had largely been completed by 2003, but it took another year to secure the necessary approvals and sign an agreement with freight operator Pacific National over the use of its line into Castlemaine station, which is still in regular passenger use today.[9] The section of line finally opened on December 19, 2004, approximately a year behind schedule.[10] As of April 2006, the Society has shown no plans to restore the dismantled Shelbourne extension in the foreseeable future.

Current operations

The railway now operates two return trips to Castlemaine on Sundays, Wednesdays & public holidays. Steam locomotives operate most services, although the Society also operates diesels during days of total fire ban or when steam is not available.

Steam locomotive K160

Locomotives

K160: Operational, stand-by unit only pending maintenance. Owned by Victorian Goldfields Railway.

J515: Non-Operational due to failing boiler exam, on long term loan from Seymour Railway Heritage Centre.

J541: Operational, leased from a private owner.

J549: Non-Operational due to major overhaul, owned by the Victorian Goldfields Railway.

Y133: Operational, on long term loan from Seymour Railway Heritage Centre.

F212: Limited use within Maldon yard, owned by the Victorian Goldfields Railway.

T333: On long term loan to Seymour Railway Heritage Centre, based at their Seymour depot.

RM61(railmotor): Operational, owned by the Victorian Goldfields Railway. 15 June 2010 saw this unit transferred to The South Gippsland Railway on a long term loan. It is rumoured that Seymour Railway Heritage Centre's DRC unit will eventually transfer to Maldon to replace RM61.

Highlights

29 March 1986: First section of railway reopened between Maldon station and the Bendigo Road crossing (approx. 1 km). The reopening train consist being K160-2BCE-42BU-16BCPL.

1987: Diesel locomotives F212 and T333 acquired from VicRail. F212 was transferred from Melbourne to Maldon by road. T333 was railed from Bendigo to Maldon.

31 March 1991: Locomotive J549 makes its first move under steam and returned to service later that year.

9 October 1993: Railmotor 61RM officially handed over to the VGR. It had been held by a group of railmotor drivers at Spencer St prior to being railed to Bendigo and then trucked to Maldon.

25 September 1995: Locomotive D3 646 retrieved from a park outside Maryborough Railway Station.

8 June 1996: Line officially reopened to Muckleford. J549 was used on the official train.

11–12 July 1998: First heritage rail weekend, promoted as "Steamfreight '98". A variety of passenger, goods and mixed trains ran with K160, J549, T333, F212 and 61RM over the 2 days.

July 1999: "Steamfreight '99", the 2nd such event held. A similar program of operations to the previous one. Highlights included K160 being renumbered K157 on the Sunday. A triple headed goods also operated with 'K157'-T333-F212.

May 2001: "Steamfreight 2001", the 3rd such event to be held. K160, J549, T333 and 61RM operated. 2 L type sheep wagons were transferred by road from the collection at Castlemaine and restored for this event.

June 19–20, 2007: The VGR held a Steam on Show Event using J515, K160 & diesel Y133. A variety of passenger, goods and mixed trains operated from Maldon to Castlemaine on the 19th, while regular services ran as mixed trains headed by K160 and J515 on the 20th.

July 26, 2008: The VGR ran the first double-headed J Class working in thirty years, using J515 and J541.

September 6, 2008: The VGR held the first "Picnic at Muckleford" event, combined with Steam on Show. The major highlight of this event was the first triple headed steam on the VGR (and the Castlemaine-Maldon branchline) with K160-J515-J541. A variety of passenger and mixed trains ran throughout the day.

June 14, 2009: The VGR celebrated the 125th anniversary of the Castlemaine-Maldon line with a special train from Castlemaine hauled by doubled headed J class. Steamrail Victoria operated double headed R class from Melbourne as part of the festivities.

December 19, 2010: VGR owned E cars 2BCE, 20BE, 15BE & 18AE, were transferred from Maldon to Seymour for restoration. [11]

Maldon Station fire

During the late hours of **October 20, 2009** a fire ripped through Maldon station building causing major damage and loss of equipment. Not to be brought down by the situation, The VGR ran regular services as usual on the 21st hauled by K160. At the time of this addition, the major brick work structure of the station is deemed safe to rebuild on, although the remaining damaged roof will have to be demolished, plans are underway to return the station building to its former glory as time and finances permit. Contract work on the roof through VicTrack commenced on 4th February 2011.

List of stations

See also

- Tourist and Heritage Railways Act

References

[1] "VR History" (http://www.victorianrailways.net/vr history/history.html). www.victorianrailways.net. 2007-02-02. . Retrieved 2007-01-20.

[2] "VGR Timechart 1881-1890" (http://www.vgr.com.au/timechart/ timechart1881-1890.htm). www.vgr.com.au. 2005-02-27. . Retrieved 2007-01-17.

[3] "VGR Timechart 1921-1940" (http://www.vgr.com.au/timechart/ timechart1921-1940.htm). www.vgr.com.au. 2005-02-27. . Retrieved 2007-01-17.

[4] "VGR Timechart 1941-1960" (http://www.vgr.com.au/history.html). www.vgr.com.au. 2005-02-27. . Retrieved 2007-01-17.

[5] "History & Preservation" (http://www.vgr.com.au/history.html). www.vgr.com.au. 2006-09-16. . Retrieved 2007-01-17.

[6] "Mount Alexander Shire Annual Report 2004 Part 1" (http://www. mountalexander.vic.gov.au/Files/AnnualReport2004Part1.pdf) (pdf). pp. 34. . Retrieved 2007-04-30.

[7] Victorian Government (2005-03-07). "PREMIER AND SCHOOL STUDENTS SEE THE SIGHTS OF MT ALEXANDER SHIRE VIA STEAM TRAIN" (http://www. legislation.vic.gov.au/domino/Web_Notes/newmedia.nsf/ 798c8b072d117a01ca256c8c0019bb01/ 297c4b491af148dcca256fbf000f8860!OpenDocument) (pdf). Press release. . Retrieved 2007-04-30.

[8] "$20 Million For 13,000 New Work For The Dole Places" (http://mediacentre. dewr.gov.au/mediacentre/AllReleases/1998/December/ 20MillionFor13000NewWorkForTheDolePlaces.htm). Commonwealth of Australia; Department of Employment and Workplace Relations. 1998-12-18. . Retrieved 2007-04-30.

[9] Catharine Munro (2004-12-20). "Goldfields railway trundles into history" (http:// www.theage.com.au/news/National/Goldfields-railway-trundles-into-history/ 2004/12/19/1103391638716.html). *The Age*. . Retrieved 2007-01-17.

[10] "MUCKLEFORD - CASTLEMAINE RESTORATION" (http://web.archive.org/web/20041211225512/www.vgr.com.au/ muckleford-castlemaine.html). www.vgr.com.au. 2004-10-20. Archived from the original (http://www.vgr.com.au/ muckleford-castlemaine.html) on 2004-11-28. . Retrieved 2007-01-20.

[11] http://www.railpage.com.au/f-t11351359-0-asc-s135.htm

External links

- Victorian Goldfields Railway (http://www.vgr.com.au) Official VGR website
- Mark Parker's Victorian Goldfields Railway Photos (http://www. marksue.bizmail.com.au/vgr/vgr.htm) Various photographs of the Muckleford to Castlemaine restoration, 2002–2004
- Railpix Australia (VIC) Victorian Goldfields Rwy (http://railpix. railmedia.com.au/index.php?uid=1618) Photographs of VGR operations from 1998

Location of Shelbourne line on VR system

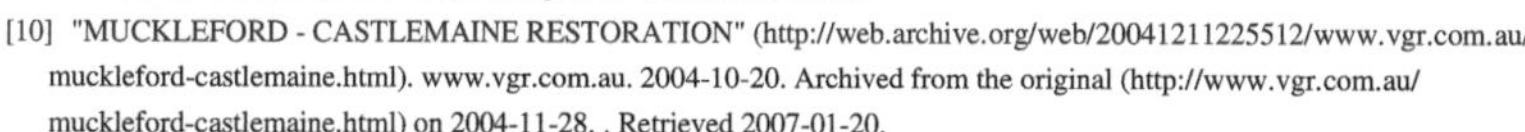

J 515 heading for Maldon station, as seen from the fireman's side window of the cab

Maldon railway station

Bendigo railway line

<table>
<tr><td colspan="2" align="center">Bendigo railway line, Victoria</td></tr>
<tr><td colspan="2">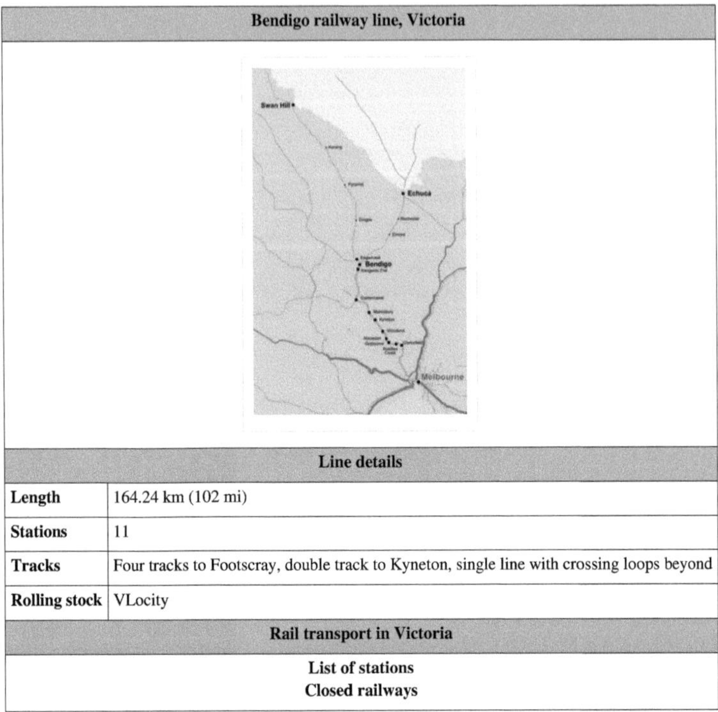</td></tr>
<tr><td colspan="2" align="center">Line details</td></tr>
<tr><td>Length</td><td>164.24 km (102 mi)</td></tr>
<tr><td>Stations</td><td>11</td></tr>
<tr><td>Tracks</td><td>Four tracks to Footscray, double track to Kyneton, single line with crossing loops beyond</td></tr>
<tr><td>Rolling stock</td><td>VLocity</td></tr>
<tr><td colspan="2" align="center">Rail transport in Victoria</td></tr>
<tr><td colspan="2" align="center">List of stations
Closed railways</td></tr>
</table>

The **Bendigo railway line** is a regional railway in Victoria, Australia. The electrified part of it within metropolitan Melbourne is the Sydenham line. It has 11 stations. The line was upgraded as part of the Regional Fast Rail project between 2005 and 2006.

History

Originally called the *Melbourne and Murray River Railway* it was constructed by the Victorian Railways Department under the supervision of Engineer in Chief George Christian Darbyshire and completed under Thomas Higginbotham. The line originating from Spencer Street Station reached Sunbury in 1859. It had reached Woodend and Kyneton by 1861 and Castlemaine in 1862. It opened in Bendigo on Thursday, 30 October 1862.[1]

Branch lines

The Lancefield branch line was opened from Clarkefield (north of Sunbury) to Lancefield in 1881, and extended to Kilmore in 1892 to connect with the Heathcote railway line. This line was completely closed by 1956.

The Daylesford branch line was opened from Carlsruhe (between Woodend and Kyneton) to Daylesford in 1880. This line was closed in 1978. Part of this line, between Daylesford and Bullarto is now operated by the Daylesford Spa Country Railway as a tourist railway. This line was connected with a line from Ballarat in 1887.

A branch line was built between Redesdale Junction (north of Kyneton) and Redesdale by 1900. It was closed in the 1950s.

The Maldon branch line was opened from Castlemaine to Maldon in 1884 and Shelbourne in 1891, although the railway had originally been planned to run to Laanecoorie. The line from Maldon to Shelbourne was closed in 1969 due to bush fire damage. The branch line closed in 1976.[2] The Victorian Goldfields Railway has restored the line and operates trains between Castlemaine and Maldon.

A branch line was built from Bendigo to Heathcote in 1888 and connected to a line from Kilmore in 1890. The Bendigo – Heathcote line closed in 1958. The Heathcote Junction to Heathcote branch was closed in November 1968.

Echuca line

The Bendigo line was extended to Elmore and Echuca in 1864 and across the Murray River to connect with the private Deniliquin and Moama Railway Company from Moama to Barnes and Deniliquin in 1876. This company was taken over by Victorian Railways in 1923.

A branch line was built from Elmore to Cohuna in 1910 and it was closed in the 1980s.

A branch line was built from Barnes to Moulamein and Balranald in 1926. The Moulamein – Balranald section was closed in the 1980s.

In 1996 the passenger service to Echuca was reinstated for the first time since 1983, when a twice-weekly service from Bendigo was started. Since 2007, there is one train to/from Melbourne on weekdays and two on weekends, with the train speed between Bendigo and Echuca limited to 80 km/h.

Swan Hill line

The Swan Hill line was extended north from Eaglehawk (just north of Bendigo on the line to Inglewood) in 1882, reaching Swan Hill in 1890. It remains in use today.

Robinvale line

The Robinvale line was opened from Bendigo to Inglewood in 1876, Korong Vale in 1882, Boort in 1883, Quambatook in 1894, Ultima in 1900, Chillingolah in 1909, Manangatang in 1914, Annuello in 1921 and Robinvale in 1924. This line currently only handles grain trains.

Victorian Railways commenced construction of a railway to Koorakee and Lette in New South Wales in 1924, but this railway was never completed. The Murray River bridge between Robinvale and Euston was instead converted to a road bridge. A new road bridge is currently being built to replace it and the old bridge will be demolished on its completion. A short branch line was built from Wedderburn Junction (south of Korong Vale) to Wedderburn in the 1880s and it was closed in the 1980s.

There is currently no passenger service on this line.

Kulwin line

The Kulwin line was opened from Korong Vale to Wycheproof in 1883, Sea Lake in 1895, Nandaly in 1914, Mittyack in 1919 and Kulwin in 1919.

This line currently only handles grain trains. Until late 2006, rural rail network lessee Pacific National had mothballed the Mittyack to Kulwin section but this has been recently re-opened to traffic despite the poor grain harvest. There has not been passenger service on this line since before 1984.

Line guide

Regular V/Line passenger services operate along the Bendigo line. Some services continue beyond Bendigo as the Swan Hill and Echuca lines. Suburban Metro Trains Melbourne services operate to Watergardens as the Sydenham line, the current limit of electrification on the line.

There is double track as far as Kyneton station, with a single track line continuing beyond. The line beyond Kyneton was previously double track prior to the Regional Fast Rail project in 2005.[3]

During peak hour some services terminate at Sunbury or Kyneton stations. Passengers on the inner section of the line are permitted to use Metcard tickets to access the services, with this section marked as the Sunbury line on suburban network maps.

Bold stations are termini, *italic* stations are staffed at least part time (this has been confirmed).

The portal of the *Big Hill* railway tunnel, 390 metres long

References

[1] NLA Australian Newspapers—The Courier (Brisbane, Qld.: 1861 – 1864)—Thursday 30 October 1862 (http://trove.nla.gov.au/ndp/del/article/4608900)

[2] "Chart of Events" (http://www.vgr.com.au/timechart/timechart1988.htm). Victorian Goldfields Railway (http://www.vgr.com.au). . Retrieved 2006-06-09.

[3] http://www.legislation.vic.gov.au/domino/Web_Notes/newmedia.nsf/b0222c68d27626e2ca256c8c001a3d2d/cc4932f92392b214ca256de10078b4f8!OpenDocument Media Release, Minister for Public Transport, Victoria, Monday November 17, 2003

External links

* http://www.vline.com.au (http://www.vline.com.au)
* Official map (http://www.vline.com.au/maps/maps/north.html)
* Statistics and detailed schematic map (http://www.vicsig.net/index.php?page=infrastructure§ion=lineguide&line=Bendigo) at the vicsig (http://www.vicsig.net/) enthusiast website

One of many closed stations on the line, Taradale

Kyneton station

Swing gates protecting the level crossing at
Kyneton, now removed

A mix of new and old signalling at Castlemaine
station

Castlemaine, Victoria

<table>
<tr><td colspan="2" align="center">Castlemaine
Victoria</td></tr>
<tr><td colspan="2" align="center">View over central Castlemaine from the Burke and Wills Memorial Lookout</td></tr>
<tr><td colspan="2">[[File:|270px|Castlemaine is located in]]
<div style="position: absolute; z-index: 2; top: %; left: %; height: 0; width: 0; margin: 0; padding: 0;">

Castlemaine</td></tr>
<tr><td>Population:</td><td>6,797 (2006)[1]</td></tr>
<tr><td>Established:</td><td>1851</td></tr>
<tr><td>Postcode:</td><td>3450</td></tr>
<tr><td>Coordinates:</td><td>37°3′49″S 144°13′2″E</td></tr>
<tr><td>Elevation:</td><td>310.9 m (1020 ft)</td></tr>
<tr><td>Location:</td><td>• 119 km (74 mi) from Melbourne
• 38 km (24 mi) from Bendigo</td></tr>
<tr><td>LGA:</td><td>Shire of Mount Alexander</td></tr>
<tr><td>State District:</td><td>Bendigo West</td></tr>
<tr><td>Federal Division:</td><td>Bendigo</td></tr>
</table>

Mean max temp	Mean min temp	Annual rainfall
20.2 °C 68 °F	6.7 °C 44 °F	558.4 mm 22 in

Castlemaine (◀ /ˈkæsəlmeɪn/)[2] is a city in Victoria, Australia, in the Goldfields region of Victoria about 120 kilometres northwest by road from Melbourne, and about 40 kilometres from the major provincial centre of Bendigo. It is the administrative and economic centre of the Shire of Mount Alexander. The population at the 2006 Census was 6,797.[1]

It was named by chief goldfield commissioner, Captain W. Wright in honour of his Irish uncle, Viscount Castlemaine.

Castlemaine began as a gold rush boomtown in 1851, it has since become a major regional centre known for its industrial and cultural institutions including the oldest continuously operating theatre in mainland Australia, the Theatre Royal.

History

Toponomy

The first European settlers named it **Forrest Creek** and as the population grew it became known as **Mount Alexander**. The old name is still present in some place names in Victoria including the Shire of Mount Alexander and the former main road leading to it from Melbourne - Mount Alexander Road and several local institutions such the hospital and sports clubs.

In 1854, Chief goldfield commissioner, Captain W. Wright, renamed the settlement to Castlemaine in honour of his Irish uncle, Viscount Castlemaine.

Pre-history and first settlement

Castlemaine exists on the traditional lands of the Dja Dja Wurrung people, also known as the Jaara people.[3]

In September 1851, three shepherds and a bullock driver discovered gold in Specimen Gully, about 5 kilometres (3.1 mi) to the north-east. Within a month the alluvial bed of Forrest Creek was being worked with 8,000 miners on the field by the end of the year and 25,000 by March 1852.

The first Post Office opened as Forrest Creek on 1 March 1852.[4]

The Theatre Royal[5] opened in 1856 to provide entertainment for the gold diggers, with the first performance being provided by the world renowned Lola Montes and her celebrated Spider Dance.

A local government was formed on 23 April 1855 and was later to become the City of Castlemaine.[6]

In 1859, the historic Castlemaine Football Club was established, and recent evidence makes it the second oldest football club in Australia and one of the oldest football clubs in the world.

Post Gold Rush

After gold mining gradually ceased a number of other secondary industries sprang up. These included breweries, iron foundries and a woollen mill. Whilst for long Thompson's foundry, (now trading as Flowserve) was one of Castlemaine's largest employers.

Barker Street, Castlemaine in 1908

Geography

Castlemaine is nestled in a mountain valley. The urban area extends to several suburban areas, north toward Barkers Creek, west to McKenzie Hill, east to Moonlight Flat and Chewton and south to Campbells Creek.

Governance

In local government, the Castlemaine region is covered by the Shire of Mount Alexander. The council was created in 1995 as an amalgamation of a number of other municipalities in the region with the council chambers located at the Castlemaine Town Hall in central Castlemaine. Castlemaine is represented by the Castlemaine Ward.

In state politics, Castlemaine is located in the Legislative Assembly districts of Bendigo West currently held by the Australian Labor Party.[7]

In federal politics, Castlemaine is located in a single House of Representatives division – the Division of Bendigo. The Division of Bendigo has been an Australian Labor Party seat since 1998.

Economy

Castlemaine's largest industry is in manufacturing, particularly foods manufacturing and tourism.

Shire of Mount Alexander meets at Castlemaine Town Hall

The biggest employer is KR Castlemaine (formerly the Castlemaine Bacon Company established 1905), producing small goods with over 900 employees.[8]

Cultural and heritage tourism is another large industry in Castlemaine, with the historic art gallery being a major drawcard.

Heritage

Barker Street was named after William Barker, another pioneer pastoralist whose run included part of the land that is now Castlemaine. The whole eastern side of Barker Street, between Templeton Street and Lyttleton Street, has been classified by the National Trust. Adjacent the solicitors' offices is the library, built in 1857 as a mechanics' institute with additions in 1861, 1872 and 1893. Next to it is the Faulder Watson Hall which opened in 1895 and adjacent is the old telegraph office (1857). There is an information plaque. On the Lyttleton Street corner is the decorative Classical Revival post office (1873–75). It is in the form of an Italian palazzo with a central clock tower, five arched bays and strongly contrasting colouration. This structure replaced a wooden post office which was built on this same spot in 1859 when the service was transferred from the gold commissioner's camp. Over the road is the Cumberland Hotel (1884).[9]

Castlemaine Post Office

Parks and Open Space

Castlemaine has its own botanical gardens established in 1860 which are on the Victorian Heritage Register.[10] The gardens feature many exotic tree species and structures dating to the Victorian era.

The **Castlemaine Diggings National Heritage Park** is the first of its kind in Australia. It embraces goldrush relics and bushland. Home to rare and threatened species of both flora and fauna it offers opportunities for bush walking, bird watching, wildlife monitoring and study while providing a bush setting for the township.

Art gallery and museum

Founded in 1913, the Castlemaine Art Gallery and Historical Museum has acquired a collection of Australian art works and historical items from the district's past. The 1931 art deco building is noted for its elegant design and is Heritage listed. The building has been extended a number of times.[11]

The art deco Castlemaine Art Gallery and Historical Museum, built in 1931

The Gallery has always specialised in Australian art.[12] Its particular strength is in major works of the late 19th century, the Golden Period of Australian painting, and the Edwardian era. Traditional landscape painting is a feature of the collection. More contemporary artists are also well represented.

Earlier artists include Louis Buvelot, Fred McCubbin, Tom Roberts, Arthur Streeton, Walter Withers, David Davies, Rupert Bunny, Max Meldrum, John Russell, Hugh Ramsay, Clarice Beckett, Arthur Lindsay and John Longstaff.

Modernists include Margaret Preston, Roland Wakelin, Russell Drysdale, Fred Williams, John Brack, Albert Tucker, John Perceval, Clifton Pugh, Lloyd Rees and Roger Kemp. More contemporary painters include Rick Amor, John Dent, Ray Crooke, Peter Benjamin Graham, Robert Jacks, Jeffrey Smart, Ian Armstrong, Paul Cavell and Brian Dunlop.The Gallery collects photographic images of Australian artists by Australian photographers and has built up a collection by photographers as Max Dupain, David Moore, Richard Beck, and Olive Cotton.

The Museum houses a collection of historical artworks, journals, photographs, costumes, and items specifically relating to the history of the Mount Alexander district.

Culture

The Theatre Royal claims to be the oldest continuously operating theatre in mainland Australia. It hosts films (including several world and Australian premieres), concerts and functions.

Castlemaine State Festival

For the past thirty years Castlemaine has biennially been the home of The Castlemaine State Festival, one of Victoria's premier regional arts events. The Festival, usually held in late March, has on offer over 130 events, many of which are free, with a particular emphasis on outdoor events, visual arts, music and theatre. It has also attracted internationally and nationally renowned performers, including The Melbourne Symphony Orchestra.

Sport

Australian rules football is popular with the Castlemaine Football Club competing in the Bendigo Football League.[13] The club is acknowledged as being the second oldest in Australia.

Castlemaine is also the self proclaimed Hot Rod centre of Australia with many small businesses catering to this popular form of motor sport at a national level. Fine examples of the cars can be seen on show days and rod runs throughout the year. There are plans for a permanent Hot Rod Centre with many community facilities currently being developed.

On Wesley Hill, just out of Castlemaine, the Castlemaine Sporting complex is situated, which hosts a range of sports, from basketball and netball, to badminton. Tennis is played on the local courts.

Castlemaine has many cricketing teams in their league.

Golfers play at the Castlemaine Golf Club on Newstead Road [14] or at the course of the Mount Alexander Golf Club on Wimble Street.[15]

The area is also renowned for its mountain bike trails.

In Popular Culture

Castlemaine XXXX was named after it. HMAS Castlemaine was also named after it.

Infrastructure

Health

A large hospital and geriatric centre (Mount Alexander Hospital) and a correctional facility is located on the eastern outskirts at (Wesley Hill).

Transport

Castlemaine is at the junction of several main roads including the Pyrenees Highway running west connecting it to Maryborough and east toward Elphinstone, the Midland Highway running north connecting it to Bendigo and south connecting it to Daylesford and Maldon-Castlemaine Road, running north west toward Maldon.

Castlemaine Railway Station

Rail services operate out of Castlemaine railway station which is on the Bendigo railway line. V/Line operates VLocity services to Melbourne's Southern Cross Station, the fastest weekday express taking 65 minutes. Travel to Bendigo by train takes a minimum 18 minutes. Victorian Goldfields Railway also operates a tourist railway running old steam and diesel engines from Castlemaine station along the route to Maldon via Muckleford.

Castlemaine Bus Lines provides suburban bus services to Chewton and Campbells Creek as well as intercity services to Maldon and Bendigo. The local taxi service is run by Castlemaine Taxis.

Media

The main newspaper is known as the **Castlemaine Mail** which began as the Mount Alexander Mail in 1854.

Prominent people from Castlemaine

- Ron Barassi - Australian rules footballer from Guildford within the shire
- Robert O'Hara Burke – leader of the Burke and Wills expedition, stationed in Castlemaine as police superintendent from 1858 to 1859
- Frank Laver – Test cricketer
- Sir Harry Lawson – Premier of Victoria
- Dustin Martin - Australian Rules footballer for Richmond, picked 3rd overall in 2009 AFL Draft
- Sir James Whiteside McCay - lawyer, teacher, politician, soldier
- Frank McEncroe - creator of the Chiko Roll
- Steven Oliver – former Carlton footballer and candidate in the 2010 Victorian state election
- Sir James Patterson – Premier of Victoria
- Frank Tate - Victoria's first Director of Education, a position he held for 26 years

Bust of Harry Lawson, Premier of Victoria 1918–1923

See also

- HM Prison Loddon

References

[1] Australian Bureau of Statistics (25 October 2007). "Community Profile Series : Castlemaine" (http://www.censusdata.abs.gov.au/ABSNavigation/prenav/ProductSelect?newproducttype=Community+Profiles&collection=Census&period=2006&areacode=SSC25327&breadcrumb=LP¤taction=201&action=401). *2006 Census of Population and Housing*. . Retrieved 2007-06-30.

[2] *Macquarie Dictionary, Fourth Edition* (2005). Melbourne, The Macquarie Library Pty Ltd. ISBN 1-876429-14-3

[3] http://en.wikipedia.org/wiki/Dja_Dja_Wurrung

[4] Premier Postal History, *Post Office List* (https://www.premierpostal.com/cgi-bin/wsProd.sh/Viewpocdwrapper.p?SortBy=VIC&country=), , retrieved 2008-04-11

[5] The Theatre Royal Castlemaine (http://www.theatreroyal.info/)

[6] *Victorian Municipal Directory*. Brunswick: Arnall & Jackson. 1992. pp. 332. Accessed at State Library of Victoria, La Trobe Reading Room.

[7] "State Election 2006 Results: Electorate swings" (http://web.archive.org/web/20071130000929/http://www.vec.vic.gov.au/electorateswing.html). *Victorian Electoral Commission website*. Archived from the original (http://www.vec.vic.gov.au/electorateswing.html) on 30 November 2007. . Retrieved 16 December 2007.

[8] Is Don. Is good for Castlemaine (http://www.bendigoadvertiser.com.au/news/local/news/general/is-don-is-good-for-castlemaine/1231006.aspx) Bendigo Advertiser. 31 Jul, 2008

[9] "Castlemaine" (http://www.smh.com.au/news/Victoria/Castlemaine/2005/02/17/1108500206304.html). *The Sydney Morning Herald*. 2008-09-15. .

[10] http://vhd.heritage.vic.gov.au/places/heritage/1791

[11] Castlemaine Art Gallery & Historical Museum (http://www.castlemainegallery.com/)

[12] http://www.collectionsaustralia.net/venues_item/59

[13] Full Points Footy, *Castlemaine* (http://www.fullpointsfooty.net/Castlemaine.htm), , retrieved 2008-07-25

[14] Golf Select, *Castlemaine* (http://www.golfselect.com.au/armchair/courseView.aspx?course_id=946), , retrieved 2009-05-11

[15] Golf Select, *Mount Alexander* (http://www.golfselect.com.au/armchair/courseView.aspx?course_id=967), , retrieved 2009-05-11

External links

- Castlemaine Maldon: Visitor Information (http://www.maldoncastlemaine.com)
- Castlemaine Historical Society (http://www.castlemainehistoricalsociety.com)
- Art Trails: Castlemaine Art Gallery " Historical Museum (http://amol.org.au/art_trails/castlemaine)

Muckleford railway station

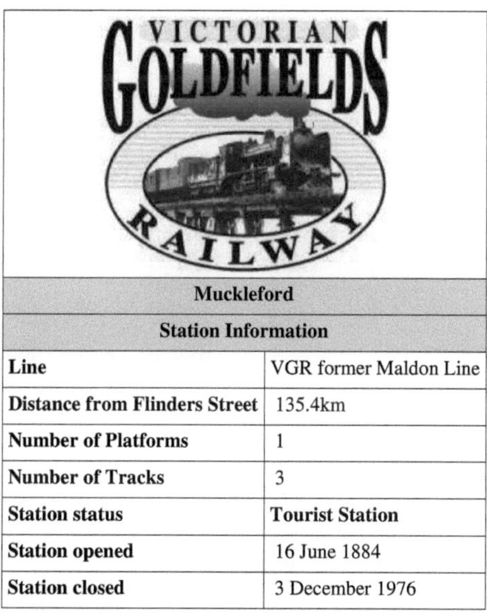

Muckleford	
Station Information	
Line	VGR former Maldon Line
Distance from Flinders Street	135.4km
Number of Platforms	1
Number of Tracks	3
Station status	**Tourist Station**
Station opened	16 June 1884
Station closed	3 December 1976

Muckleford is a railway station on the Maldon branch line off the main Echuca, Swan Hill lines in Victoria, Australia. The station was originally opened on 16 June 1884 and was closed to passenger services on 6 January 1941.[1] After this date, the line was used only for goods traffic until closure on 3 December 1976.[2]

Muckleford station was opened for tourist services in 1996 after the section of line between Maldon and Muckleford had been restored.

Station Navigation
Victorian Goldfields Railway
Previous Station Castlemaine	Maldon Next Station
Entire line

References

[1] "VGR Timechart 1941-1960" (http://www.vgr.com.au/history.html). www.vgr.com.au. 2005-02-27. . Retrieved 2007-01-17.

[2] "History & Preservation" (http://www.vgr.com.au/history.html). www.vgr.com.au. 2006-09-16. . Retrieved 2007-01-17.

Rail transport

BNSF Railway freight service in the United States

German InterCityExpress

Rail transport
Operations
Track
Maintenance
High-speed
Gauge
Stations
Trains
Locomotives
Rolling stock
Railways
History
Attractions
Terminology
By country
Accidents
Modelling

Rail transport is a means of conveyance of passengers and goods by way of wheeled vehicles running on rail tracks. In contrast to road transport, where vehicles merely run on a prepared surface, rail vehicles are also directionally guided by the tracks they run on. Track usually consists of steel rails installed on sleepers/ties and ballast, on which the rolling stock, usually fitted with metal wheels, moves. However, other variations are also possible, such as slab track where the rails are fastened to a concrete foundation resting on a prepared subsurface.

Rolling stock in railway transport systems generally has lower frictional resistance when compared with highway vehicles, and the passenger and freight cars (carriages and wagons) can be coupled into longer trains. The operation is carried out by a railway company, providing transport between train stations or freight customer facilities. Power is provided by locomotives which either draw electrical power from a railway electrification system or produce their own power, usually by diesel engines. Most tracks are accompanied by a signalling system. Railways are a safe land transport system when compared to other forms of transport.[1] Railway transport is capable of high levels of passenger and cargo utilization and energy efficiency, but is often less flexible and more capital-intensive than highway transport is, when lower traffic levels are considered.

The oldest, man-hauled railways date to the 6th century B.C, with Periander, one of the Seven Sages of Greece, credited with its invention. With the British development of the steam engine, it was possible to construct mainline railways, which were a key component of the industrial revolution. Also, railways reduced the costs of shipping, and allowed for fewer lost goods. The change from canals to railways allowed for "national markets" in which prices varied very little from city to city. Studies have shown that the invention and development of the railway in Europe was one of the most important technological inventions of the late 19th century for the United States, without which, GDP would have been lower by 7.0% in 1890. In the 1880s, electrified trains were introduced, and also the first tramways and rapid transit systems came into being. Starting during the 1940s, the non-electrified railways in most countries had their steam locomotives replaced by diesel-electric locomotives, with the process being almost complete by 2000. During the 1960s, electrified high-speed railway systems were introduced in Japan and a few other countries. Other forms of guided ground transport outside the traditional railway definitions, such as monorail or maglev, have been tried but have seen limited use.

History

Pre-steam

The earliest evidence of a railway was a 6-kilometre (3.7 mi) Diolkos wagonway, which transported boats across the Corinth isthmus in Greece during the 6th century BC. Trucks pushed by slaves ran in grooves in limestone, which provided the track element. The Diolkos ran for over 600 years.[2]

Railways began reappearing in Europe after the Dark Ages. The earliest known record of a railway in Europe from this period is a stained-glass window in the Minster of Freiburg im Breisgau in Germany, dating from around 1350.[3] In 1515, Cardinal Matthäus Lang wrote a description of the Reisszug, a funicular railway at the Hohensalzburg Castle in Austria. The line originally used

Horsecar in Brno, Czech Republic

wooden rails and a hemp haulage rope, and was operated by human or animal power. The line still exists, albeit in updated form, and is probably one of the oldest railway still to operate.[4] [5]

By 1550, narrow gauge railways with wooden rails were common in mines in Europe.[6] By the 17th century, wooden wagonways were common in the United Kingdom for transporting coal from mines to canal wharfs for transshipment to boats. The world's oldest working railway, built in 1758, is the Middleton Railway in Leeds. In

1764, the first gravity railroad in the United States was built in Lewiston, New York.[7] The first permanent tramway was the Leiper Railroad in 1810.[8]

The first iron plate rail way made with cast iron plates on top of wooden rails, was taken into use in 1768.[9] This allowed a variation of gauge to be used. At first only balloon loops could be used for turning, but later, movable points were taken into use that allowed for switching.[10] From the 1790s, iron edge rails began to appear in the United Kingdom.[11] In 1803, William Jessop opened the Surrey Iron Railway in south London, arguably the world's first horse-drawn public railway.[12] The invention of the wrought iron rail by John Birkinshaw in 1820 allowed the short, brittle, and often uneven, cast iron rails to be extended to 15 feet (4.6 m) lengths.[13] These were succeeded by steel in 1857.[11]

The railroad era in the United States began in 1830 when Peter Cooper's locomotive, *Tom Thumb*, first steamed along 13 miles (21 km) of Baltimore and Ohio railroad track.[14] In 1833 the nation's second railroad ran 136 miles (219 km) from Charleston to Hamburg in South Carolina.[15] Not until the 1850s, though, did railroads offer long distance service at reasonable rates. A journey from Philadelphia to Charleston involved eight different gauges, which meant that passengers and freight had to change trains seven times. Only at places like Bowling Green, Kentucky, the railroads were connected to one another.

Age of steam

The development of the steam engine spurred ideas for mobile steam locomotives that could haul trains on tracks. The first was patented by James Watt in 1769 and revised in 1782, but these heavy low-pressure engines were not suitable for use in locomotives. In 1804, using high-pressure steam, Richard Trevithick demonstrated the first locomotive-hauled train in Merthyr Tydfil, United Kingdom.[16] [17] Accompanied with Andrew Vivian, it ran with mixed success,[18] breaking

A British steam locomotive-hauled train

some of the brittle cast-iron plates.[19] Two years later, the first passenger horse-drawn railway was opened nearby between Swansea and Mumbles.[20]

In 1811, John Blenkinsop designed the first successful and practical railway locomotive[21] —a rack railway worked by a steam locomotive between Middleton Colliery and Leeds on the Middleton Railway. The locomotive, *The Salamanca*, was built the following year.[22] :20 In 1825, George Stephenson built the *Locomotion* for the Stockton and Darlington Railway, north east England, which was the first public steam railway in the world. In 1829, he built *The Rocket* which was entered in and won the Rainhill Trials. This success led to Stephenson establishing his company as the pre-eminent builder of steam locomotives used on railways in the United Kingdom, the United States and much of Europe.[22] :24–30

In 1830, the first intercity railway, the Liverpool and Manchester Railway, opened. The gauge was that used for the early wagonways and had been adopted for the Stockton and Darlington Railway.[23] The 1435 mm (4 ft 8 $\frac{1}{2}$ in) width became known as the international standard gauge, used by about 60% of the world's railways. This spurred the spread of rail transport outside the UK. The Baltimore and Ohio that opened in 1830 was the first to evolve from a single line to a network in the United States.[24] By 1831, a steam railway connected Albany and Schenectady, New York, a distance of 16 miles, which was covered in 40 minutes.[25] In 1867, the first elevated railway was built in New York. The symbolically important first transcontinental railway was completed in 1869.[26]

Electrification and dieselisation

Experiments with electrical railways were started by Robert Davidson in 1838. He completed a battery-powered carriage capable of 6.4 km/h (4 mph). The Giant's Causeway Tramway was the first to use electricity fed to the trains en-route, using a third rail, when it opened in 1883. Overhead wires were taken into use in 1888. At first, this was taken into use on tramways that, until then, had been horse-drawn tramcars. The first conventional electrified railway was the Roslag Line in Sweden. During the 1890s, many large cities, such as London, Paris and New York used the new technology to build rapid transit for urban commuting. In smaller cities, tramways became common and were often the only mode of public transport until the introduction of buses in the 1920s. In North America, interurbans became a common mode to reach suburban areas. At first, all electric railways used direct current but, in 1904, the Spubeital Line in Austria opened with alternating current.[27]

Elevated section of the Chicago 'L'

Steam locomotives require large pools of labour to clean, load, maintain and run. After World War II, dramatically increased labour costs in developed countries made steam an increasingly costly form of motive power. At the same time, the war had forced improvements in internal combustion engine technology that made diesel locomotives cheaper and more powerful. This caused many railway companies to initiate programmes to convert all unelectrified sections from steam to diesel locomotion.

Luas in Dublin, Ireland

Following the large-scale construction of motorways after the war, rail transport became less popular for commuting and air transport started taking large market shares from long-haul passenger trains. Most tramways were either replaced by rapid transits or buses, while high transshipment costs caused short-haul freight trains to become uncompetitive. The 1973 oil crisis led to a change of mind set and most tram systems that had survived into the 1970s remain today. At the same time, containerization allowed freight trains to become more competitive and participate in intermodal freight transport. With the 1964 introduction of the Shinkansen high-speed rail in Japan, trains could again have a dominant position on intercity travel. During the 1970s, the introduction of automated rapid transit systems allowed cheaper operation. The 1990s saw an increased focus on accessibility and low-floor trains. Many tramways have been upgraded to light rail and many cities that closed their old tramways have reopened new light railway systems.

Trains

A train is a connected series of rail vehicles that move along the track. Propulsion for the train is provided by a separate locomotive or from individual motors in self-propelled multiple units. Most trains carry a revenue load, although non-revenue cars exist for the railway's own use, such as for maintenance-of-way purposes. The engine driver controls the locomotive or other power cars, although people movers and some rapid transits are driverless.

Haulage

Traditionally, trains are pulled using a locomotive. This involved a single or multiple powered vehicles being located at the front of the train and providing sufficient adhesion to haul the weight of the full train. This remains dominant for freight trains and is often used for passenger trains. A push-pull train has the end passenger car equipped with a driver's cab so the engine driver can remotely control the locomotive. This allows one of the locomotive-hauled trains drawbacks to be removed, since the locomotive need not be moved to the end of the train each time the train changes direction. A railroad car is a vehicle used for the haulage of either passengers or freight.

Russian 2TE10U diesel locomotive

A multiple unit has powered wheels throughout the whole train. These are used for rapid transit and tram systems, as well as many both short- and long-haul passenger trains. A railcar is a single, self-powered car. Multiple units have a driver's cab at each end of the unit and were developed following the ability to build electric motors and engines small enough to build under the coach. There are only a few freight multiple units, most of which are high-speed post trains.

Motive power

Steam locomotives are locomotives with a steam engine that provides adhesion. Coal, petroleum, or wood is burned in a firebox. The heat boils water in the fire-tube boiler to create pressurized steam. The steam travels through the smokebox before leaving via the chimney. In the process, it powers a piston that transmits power directly through a connecting rod (US: main rod) and a crankpin (US: wristpin) on the driving wheel (US main driver) or to a crank on a driving axle. Steam locomotives have been phased out in most parts of the world for economical and safety reasons although many are preserved in working order by heritage railways.

A RegioSwinger multiple unit of the Croatian Railways

Electric locomotives draw power from a stationary source via an overhead wire or third rail. Some also or instead use a battery. A transformer in the locomotive converts the high voltage, low current power to low voltage, high current used in the electric motors that power the wheels. Modern locomotives use three-phase AC induction motors. Electric locomotives are the most powerful traction.

They are also the cheapest to run and provide less noise and no local air pollution. However, they require high capital investments both for the overhead lines and the supporting infrastructure. Accordingly, electric traction is used on urban systems, lines with high traffic and for high-speed rail.

Diesel locomotives use a diesel engine as the prime mover. The energy transmission may be either diesel-electric, diesel-mechanical or diesel-hydraulic but diesel-electric is dominant. Electro-diesel locomotives are built to run as diesel-electric on unelectrified sections and as electric locomotives on electrified sections.

Alternative methods of motive power include magnetic levitation, horse-drawn, cable, gravity, pneumatics and gas turbine.

Passenger trains

A passenger train travels between stations where passengers may embark and disembark. The oversight of the train is the duty of a guard/train manager. Passenger trains are part of public transport and often make up the stem of the service, with buses feeding to stations.

Intercity trains are long-haul trains that operate with few stops between cities. Trains typically have amenities such as a dining car. Some lines also provide over-night services with sleeping cars. Some long-haul trains been given a specific name. Regional trains are medium distance trains that connect cities with outlying, surrounding areas, or provide a regional service, making

Interior view of the top deck of a VR InterCity2 double-deck carriage

more stops and having lower speeds. Commuter trains serve suburbs of urban areas, providing a daily commuting service. Airport rail links provide quick access from city centres to airports.

Rapid transit is built in large cities and has the highest capacity of any passenger transport system. It is grade separated and commonly built underground or elevated. At street level, smaller trams can be used. Light rails are upgraded trams that have step-free access, their own right-of-way and sometimes sections underground. Monorail systems operate as elevated, medium capacity systems. A people mover is a driverless, grade-separated train that serves only a few stations, as a shuttle.

High-speed rail operates at much higher speeds than conventional railways, the limit being regarded at 200 to 320 km/h. High-speed trains are used mostly for long-haul service and most systems are in Western Europe and East Asia. The speed record is 574.8 km/h (357.2 mph), set by a modified French TGV.[28] [29] Magnetic levitation trains such as the Shanghai airport train use under-riding magnets which attract themselves upward towards the underside of a guideway and this line has achieved somewhat higher peak speeds in day-to-day operation than conventional high-speed railways, although only over short distances.

Freight train

A freight train hauls cargo using freight cars specialized for the type of goods. Freight trains are very efficient, with economy of scale and high energy efficiency. However, their use can be reduced by lack of flexibility, if there is need of transshipment at both ends of the trip due to lack of tracks to the points of pick-up and delivery. Authorities often encourage the use of cargo rail transport due to its environmental profile.

Bulk cargo of minerals

Container trains have become the dominant type in the US for non-bulk haulage. Containers can easily be transshipped to other modes, such as ships and trucks, using cranes. This has succeeded the boxcar (wagon-load), where the cargo had to be loaded and unloaded into the train manually. In Europe, the sliding wall wagon has largely superseded the ordinary covered wagons. Other types of cars include refrigerator cars, stock cars for livestock and autoracks for road vehicles. When rail is combined with road transport, a roadrailer will allow trailers to be driven onto the train, allowing for easy transition between road and rail.

Bulk handling represents a key advantage for rail transport. Low or even zero transshipment costs combined with energy efficiency and low inventory costs allow trains to handle bulk much cheaper than by road. Typical bulk cargo includes coal, ore, grains and liquids. Bulk is transported in open-topped cars and tank cars.

Infrastructure

Right of way

Railway tracks are laid upon land owned or leased by the railway company. Owing to the desirability of maintaining modest grades, rails will often be laid in circuitous routes in hilly or mountainous terrain. Route length and grade requirements can be reduced by the use of alternating cuttings, bridges and tunnels—all of which can greatly increase the capital expenditures required to develop a right of way, while significantly reducing operating costs and allowing higher speeds on longer radius curves. In densely urbanized areas, railways are sometimes laid in tunnels to minimize the effects on existing properties.

Trackage

Track consists of two parallel steel rails, anchored perpendicular to members called ties (sleepers) of timber, concrete, steel, or plastic to maintain a consistent distance apart, or gauge. The track guides the conical, flanged wheels, keeping the cars on the track without active steering and therefore allowing trains to be much longer than road vehicles. The rails and ties are usually placed on a foundation made of compressed earth on top of which is placed a bed of ballast to distribute the load from the ties and to prevent the track from buckling as the ground settles over time under the weight of the vehicles passing above.

The ballast also serves as a means of drainage. Some more modern track in special areas is attached by direct fixation without ballast. Track may be prefabricated or assembled in place. By welding rails together to form lengths of continuous welded rail, additional wear and tear on rolling stock caused by the small surface gap at the joints between rails can be counteracted; this also makes for a quieter ride (passenger trains).

On curves the outer rail may be at a higher level than the inner rail. This is called superelevation or cant. This reduces the forces tending to displace the track and makes for a more comfortable ride for standing livestock and

Long freight train crossing the Stoney Creek viaduct on the Canadian Pacific Railway in southern British Columbia

standing or seated passengers. A given amount of superelevation will be the most effective over a limited range of speeds.

Turnouts, also known as points and switches, are the means of directing a train onto a diverging section of track. Laid similar to normal track, a point typically consists of a frog (common crossing), check rails and two switch rails. The switch rails may be moved left or right, under the control of the signalling system, to determine which path the train will follow.

Spikes in wooden ties can loosen over time, but split and rotten ties may be individually replaced with new wooden ties or concrete substitutes. Concrete ties can also develop cracks or splits, and can also be replaced individually. Should the rails settle due to soil subsidence, they can be lifted by specialized machinery and additional ballast tamped under the ties to level the rails.

Periodically, ballast must be removed and replaced with clean ballast to ensure adequate drainage. Culverts and other passages for water must be kept clear lest water is impounded by the trackbed, causing landslips. Where trackbeds are placed along rivers, additional protection is usually placed to prevent streambank erosion during times of high water. Bridges require inspection and maintenance, since they are subject to large surges of stress in a short period of time when a heavy train crosses.

Train Inspection Systems

The inspection of railway equipment is essential for the safe movement of trains. Many types of defect detectors are in use on the world's railroads. These devices utilize technologies vary from a simplistic paddle and switch to infrared and laser scanning, and even ultrasonic audio analysis. Their use has avoided many rail accidents over the 70 years they have been used.

A Hot bearing detector w/ dragging equipment unit

Signalling

Railway signalling is a system used to control railway traffic safely to prevent trains from colliding. Being guided by fixed rails with low friction, trains are uniquely susceptible to collision since they frequently operate at speeds that do not enable them to stop quickly or within the driver's sighting distance. Most forms of train control involve movement authority being passed from those responsible for each section of a rail network to the train crew. Not all methods require the use of signals, and some systems are specific to single track railways.

Great Western Railway semaphore-type signal

The signalling process is traditionally carried out in a signal box, a small building that houses the lever frame required for the signalman to operate switches and signal equipment. These are placed at various intervals along the route of a railway, controlling specified sections of track. More recent technological developments have made such operational doctrine superfluous, with the centralization of signalling operations to regional control rooms. This has been facilitated by the increased use of computers, allowing vast sections of track to be monitored from a single location. The common method of block signalling divides the track into zones guarded by combinations of block signals, operating rules, and automatic-control devices so that only one train may be in a block at any time.

Electrification

The electrification system provides electrical energy to the trains, so they can operate without a prime mover onboard. This allows lower operating costs, but requires large capital investments along the lines. Mainline and tram systems normally have overhead wires, which hang from poles along the line. Grade-separated rapid transit sometimes use a ground third rail.

Power may be fed as direct or alternating current. The most common currencies are 600 and 750 V for tram and rapid transit systems, and 1,500 and 3,000 V for mainlines. The two dominant AC systems are 15 kV AC and 25 kV AC.

Stations

A railway station serves as an area where passengers can board and alight from trains. A goods station is a yard which is exclusively used for loading and unloading cargo. Large passenger stations have at least one building providing conveniences for passengers, such as purchasing tickets and food. Smaller stations typically only consist of a platform. Early stations were sometimes built with both passenger and goods facilities.[30]

Chhatrapati Shivaji Terminus in Mumbai, India

Platforms are used to allow easy access to the trains, and are connected to each other via underpasses, footbridge and level crossings. Some large stations are built as cul-de-sac, with trains only operating out from one direction. Smaller stations normally serve local residential areas, and may have connection to feeder bus services. Large stations, in particular central stations, serve as the main public transport hub for the city, and have transfer available between rail services, and to rapid transit, tram or bus services.

Operations

Ownership

Traditionally, the infrastructure and rolling stock are owned and operated by the same company. This has often been by a national railway, while other companies have had private railways. Since the 1980s, there has been an increasing tendency to split up railway companies, with separate companies owning the stock from those owning the infrastructure, particularly in Europe, where this is required by the European Union. This has allowed open access by any train operator to any portion of the European railway network.

In the United States, railways, such as Union Pacific, are privately owned

Financing

The main source of income for railway companies is from ticket revenue (for passenger transport) and shipment fees for cargo. Discounts and monthly passes are sometimes available for frequent travellers. Freight revenue may be sold per container slot or for a whole train. Sometimes, the shipper owns the cars and only rents the haulage. For passenger transport, advertisement income can be significant.

Government may choose to give subsidies to rail operation, since rail transport has fewer externalities than other dominant modes of transport. If the railway company is state-owned, the state may simply provide direct subsidies in exchange for an increased production. If operations have been privatized, several options are available. Some countries have a system where the infrastructure is owned by a government agency or company—with open access to the tracks for any company that meets safety requirements. In such cases, the state may choose to provide the tracks free of charge, or for a fee that does not cover all costs. This is seen as analogous to the government providing free access to roads. For passenger operations, a direct subsidy may be paid to a public-owned operator, or public service obligation tender may be helt, and a time-limited contract awarded to the lowest bidder.

Safety

Main articles: List of rail accidents pre-1900; 1900–1949; 1950–1999; 2000–2009; 2010–present.

Rail transport is one of the safest forms of land travel.[31] Trains can travel at very high speed, but they are heavy, are unable to deviate from the track and require a great distance to stop. Possible accidents include derailment (jumping the track), a collision with another train or collision with an automobile or other vehicle at level crossings. The latter accounts for the majority of rail accidents and casualties. The most important safety measures to prevent accidents are strict operating rules, e.g. railway signalling and gates or grade separation at crossings. Train whistles, bells or horns warn of the presence of a train, while trackside signals maintain the distances between trains.

An important element in the safety of many high-speed inter-city networks such as Japan's Shinkansen is the fact that trains only run on dedicated railway lines, without level crossings. This effectively eliminates the potential for collision with automobiles, other vehicles and pedestrians, vastly reduces the likelihood of collision with other trains and helps ensure services remain timely.

Train crash at Montparnasse Station, Paris, France, in 1895

Impact

Energy

Rail transport is an energy-efficient [32] but capital-intensive, means of mechanized land transport. The tracks provide smooth and hard surfaces on which the wheels of the train can roll with a minimum of friction. As an example, a typical modern wagon can hold up to 113 tonnes of freight on two four-wheel bogies. The contact area between each wheel and the rail is a strip no more than a few millimetres wide, which minimizes friction. The track distributes the weight of the train evenly, allowing significantly greater loads per axle and wheel than in road transport, leading to less wear and tear on the permanent way. This can save energy compared with other forms of transport, such as road transport, which depends on the friction between rubber tires and the road. Trains have a small frontal area in relation to the load they are carrying, which reduces air resistance and thus energy usage.

In addition, the presence of track guiding the wheels allows for very long trains to be pulled by one or a few engines and driven by a single operator, even around curves, which allows for economies of scale in both manpower and energy use; by contrast, in road transport, more than two articulations causes fishtailing and makes the vehicle unsafe.

Usage

Due to these benefits, rail transport is a major form of passenger and freight transport in many countries. In India, China, South Korea and Japan, many millions use trains as regular transport. It is widespread in European countries. Freight rail transport is widespread and heavily used in North America, but intercity passenger rail transport on that continent is relatively scarce outside the Northeast Corridor.[33]

Railway tracks running through Stanhope, United Kingdom

Africa and South America have some extensive networks such as in South Africa, Northern Africa and Argentina; but some railways on these continents are isolated lines. Australia has a generally sparse network befitting its population density, but has some areas with significant networks, especially in the southeast. In addition to the previously existing east-west transcontinental line in Australia, a line from north to south has been constructed. The highest railway in the world is the line to Lhasa, in Tibet,[34] partly running over permafrost territory. The western Europe region has the highest railway density in the world, and has many individual trains which operate through several countries despite technical and organizational differences in each national network.

See also

- Environmental design in rail transportation
- International Union of Railways
- List of rail transport topics
- List of railway companies
- Megaproject
- Passenger rail terminology
- Rail transport by country

References

Notes

[1] According to this source (http://www.railwatch.org.uk/backtrack/rw94/rw094p06.pdf), railways are safest on both a per-mile and per-hour basis, whereas air transport is safe only on a per-mile basis

[2] Lewis, M. J. T. "Railways in the Greek and Roman World" (http://www.sciencenews.gr/docs/diolkos.pdf) (pdf). . Retrieved 11 April 2009.

[3] Hylton, Stuart (2007). *The Grand Experiment: The Birth of the Railway Age 1820-1845*. Ian Allan Publishing.

[4] Kriechbaum, Reinhard (15 May 2004). "Die große Reise auf den Berg" (http://www.die-tagespost.de/Archiv/titel_anzeige.asp?ID=8916) (in German). *der Tagespost*. . Retrieved 22 April 2009.

[5] "Der Reiszug - Part 1 - Presentation" (http://www.funimag.com/funimag10/RESZUG01.HTM). Funimag. . Retrieved 22 April 2009.

[6] Georgius Agricola (1913). *De re metallica*. ISBN 0486600068.

[7] Porter, Peter (1914). *Landmarks of the Niagara Frontier*. The Author. ISBN 0665783477.

[8] Morlok, Edward K. (11 May 2005). "First permanent railroad in the U.S. and its connection to the University of Pennsylvania" (http://www.seas.upenn.edu/~morlok/morlokpage/transp_data.html). . Retrieved 19 September 2007.

[9] at Coalbrookdale *Railways (pt 1)* (http://www.1902encyclopedia.com/R/RAI/railway-01.html). Encyclopedia Britannica. 1902. ISBN 187252463X. . Retrieved 2011-02-15.

[10] Vaughan, A. (1997). *Railwaymen, Politics and Money*. London: John Murray. ISBN 0719557461.

[11] Marshall, John (1979). *The Guiness Book of Rail Facts & Feats*. ISBN 0-900424-56-7.

[12] "Surrey Iron Railway 200th - 26th July 2003" (http://www.stephensonloco.fsbusiness.co.uk/surreyiron.htm). *Early Railways*. Stephenson Locomotive Society. . Retrieved 19 September.

[13] Skempton, A.W. (2002). *A biographical dictionary of civil engineers in Great Britain and Ireland, John Birkinshaw* (http://books.google. com/books?id=jeOMfpYMOtYC&pg=PA59). pp. 59–60. ISBN 9780727729392. .

[14] "The History of the Tom Thumb - Peter Cooper" (http://inventors.about.com/library/inventors/bl_tom_thumb.htm). Inventors.about.com. . Retrieved 8 May 2011.

[15] Storey, Steve. "Charleston & Hamburg Railroad" (http://www.railga.com/charlhmbrg.html). Railga.com. . Retrieved 8 May 2011.

[16] "Richard Trevithick's steam locomotive" (http://www.museumwales.ac.uk/en/rhagor/article/trevithic_loco/). Museumwales.ac.uk. 15 December 2008. . Retrieved 8 May 2011.

[17] "Steam train anniversary begins" (http://news.bbc.co.uk/2/hi/uk_news/wales/3509961.stm). BBC. 21 February 2004. . Retrieved 8 May 2011. "A south Wales town has begun months of celebrations to mark the 200th anniversary of the invention of the steam locomotive. Merthyr Tydfil was the location where, on 21 February 1804, Richard Trevithick took the world into the railway age when he set one of his high-pressure steam engines on a local iron master's tram rails"

[18] Payton, Philip (2004). *Oxford Dictionary of National Biography*. Oxford University Press.

[19] Chartres, J.. "Richard Trevithick". In Cannon, John. *Oxford Companion to British History*. p. 932.

[20] "Early Days of Mumbles Railway" (http://www.bbc.co.uk/wales/southwest/sites/swansea/pages/mumbles_trainanniv.shtml). BBC. 15 February 2007. . Retrieved 19 September 2007.

[21] "John Blenkinsop" (http://www.britannica.com/EBchecked/topic/69274/John-Blenkinsop). *Encyclopædia Britannica*. . Retrieved 8 May 2011.

[22] Ellis, Hamilton (1968). *The Pictorial Encyclopedia of Railways*. Hamlyn Publishing Group.

[23] "Liverpool and Manchester" (http://www.spartacus.schoolnet.co.uk/RAliverpool.htm). . Retrieved 19 September 2007.

[24] Dilts, James D. (1996). *The Great Road: The Building of the Baltimore and Ohio, the Nation's First Railroad, 1828-1853* (http://books. google.com/?id=JjrCWPwvHzIC&lpg=PR18&dq=dilts b&o&pg=PA26#v=onepage&q=first). Palo Alto, CA: Stanford University Press. p. 26. ISBN 978-0804726290. .

[25] "The Journal of Ebenezer Mattoon Chamberlain 1832-5", *Indiana Magazine of History*, Vol. XV, September, 1919, No. 3, p.233ff.

[26] Ambrose, Stephen E. (2000). *Nothing Like It In The World; The men who built the Transcontinental Railroad 1863-1869* (http://books. google.com/?id=TZp_GT7PscIC&lpg=PP1&dq=ambrose nothing like it&pg=PP1#v=onepage&q=). Simon & Schuster. ISBN 0-684-84609-8. .

[27] Tokle, Bjørn (2003) (in Norwegian). *Communication gjennom 100 år*. Meldal: Chr. Salvesen & Chr. Thams's Communications Aktieselskab. p. 54.

[28] Associated Press (4 April 2007). "French train breaks speed record" (http://web.archive.org/web/20070407194558/http://www.cnn. com/2007/WORLD/europe/04/03/TGVspeedrecord.ap/index.html). *CNN*. Archived from the original (http://www.cnn.com/2007/ WORLD/europe/04/03/TGVspeedrecord.ap/index.html) on 7 April 2007. . Retrieved 3 April 2007.

[29] Fouquet, Helene and Viscousi, Gregory (3 April 2007). "French TGV Sets Record, Reaching 357 Miles an Hour (Update2)" (http://www. bloomberg.com/apps/news?pid=newsarchive&sid=aW23Aw20niIo&refer=europe). Bloomberg. . Retrieved 8 May 2011.

[30] "The Inception of the English Railway Station". *Architectural History* (SAHGB Publications Limited) 4: 63–76. 1961. doi:10.2307/1568245. JSTOR 1568245.

[31] U.S. Bureau of Transportation Statistics (2010) National Transportation Statistics. Table 2-1: Transportation Fatalities by Mode (http:// www.bts.gov/publications/national_transportation_statistics/html/table_02_01.html). (Report). Retrieved 2010-02-14.

[32] American Association of Railroads. "Railroad Fuel Efficiency Sets New Record" (http://www.progressiverailroading.com/news/article. asp?id=16740). . Retrieved 12 April 2009.

[33] "Public Transportation Ridership Statistics" (http://web.archive.org/web/20070815101950/http://www.apta.com/research/stats/ ridership/). American Public Transportation Association. 2007. Archived from the original (http://www.apta.com/research/stats/ridership/) on 15 August 2007. . Retrieved 10 September 2007.

[34] "New height of world's railway born in Tibet" (http://news.xinhuanet.com/english/2005-08/24/content_3397297.htm). Xinhua News Agency. 24 August 2005. . Retrieved 8 May 2011.

References

pfl:Aisebahn

Heritage railway

A **heritage railway** (United Kingdom, Australia and Canada), **preserved railway** (United Kingdom and Australia), **tourist railway** (Australia), or **tourist railroad** (United States) is a railway that is run as a tourist attraction, in some cases by volunteers, and which typically seeks to re-create or preserve railway scenes of the past (though not all tourist railways are heritage railways). See List of heritage railways.

Historic heavy and light rail

Heritage railways are usually railway lines which were once run as commercial railways, but were later no longer needed or were closed down, and were taken over or re-opened by volunteers or for-profit organisations. Many run on partial routes unconnected to the commercial railway network, run only seasonally, and charge high "entertainment" fares. For example the return fare from Porthmadog to Blaenau Ffestiniog on the 13 mile Festiniog railway is some £17.95 and between Caernarfon to Beddgelert £22.00 on the Welsh Highland Railway. As a result they are primarily, indeed exclusively, focused on serving the tourist and leisure markets, not local transportation needs. However in the 1990s and 2000s some heritage railways have professed to provide local transportation and to extend their running seasons to cater for commercial passenger traffic. In the United Kingdom, however, no heritage railways offer a year round daily or commuter service.

Typically a heritage railway will use steam locomotives and original rolling stock to create a supposed "period atmosphere", although some are now concentrating on more recent "modern image" diesel and electric traction supposedly to re-create the post-steam railway era.

The first heritage railway to be rescued and run entirely by volunteers was the Talyllyn Railway in Wales. This narrow gauge line, taken over by a group of enthusiasts in 1950, is recognised as the start of the preservation movement. There are now several hundred heritage railways in the United Kingdom. This large number is due in part to the closure of many minor lines in the 1960s under the Beeching Axe. These were relatively easy to revive. The first such standard gauge line to be preserved was the Bluebell Railway, though the Middleton Railway (which was not a victim of Beeching) had been preserved prior to this. The world's second preserved railway, and the first outside the United Kingdom, was the Puffing Billy Railway in Australia. This railway operates 24 km of track with much of the original rolling stock built as early as 1898.

A scene on a heritage railway. An ex-British Railways 4MT 2-6-4T tank engine takes on water at Bishops Lydeard station on the West Somerset Railway, Somerset, England.

The Darjeeling Himalayan Railway, India, a UNESCO World Heritage Site

the Historical Khyber Railway goes through the Khyber Pass, Pakistan

Heritage railways differ in the intensity of the service that can be offered. Some of the more successful British heritage railways, such as the Severn Valley Railway and the North Yorkshire Moors Railway, may have up to five or six steam engines working, operating a four-train service daily. The Great Central Railway is the only example of a preserved British main line that operates with a double track, and can operate over 50 trains on a busy gala timetable. Other smaller railways may run for seven-days-a-week throughout the summer with only one steam engine. However, following the privatisation of Britain's main-line railways, the line between not-for-profit heritage railways and for-profit branch lines may appear to have blurred. The Wensleydale Railway is an example of a commercial line run partly as

Stepney on the Bluebell railway, one of the first preserved in Britain

a heritage operation and partly (at least in intent, if not in reality) to provide local transportation. The Weardale Railway is a similar attempt to provide a commercial heritage line, so far with mixed success. The Severn Valley Railway has even operated a few goods trains on a commercial basis. In addition, a number of heritage lines now see regular freight operations. The Puffing Billy Railway operates a busier service than it regularly did in its pre-preservation working life.

In the 50 years since the Bluebell Railway reopened to traffic, the definition of private standard gauge railways in the United Kingdom as preserved railways has changed and evolved as the number of projects, length, operating days and function has altered. The 1970s distinction between narrow gauge, standard gauge and steam centres alone is no longer necessarily fit for purpose. The situation is further muddied by the huge variation in company structure of the ownership of the railway, its rolling stock and other assets. Unlike community railways the tourist railways in the UK are vertically integrated, although those operating in the main as charities have their charitable and non-charitable activities essentially separated for accounts purposes.

Distinction should be made heritage railways and preserved railways in C21st. A preserved railway should be defined as any railway that is not on a single recognised site; having perhaps more than 0.625miles/1 km, that runs from A to B, having one or more stations and the railway itself is generally accepted as the main attraction. The term heritage railway should reserved for lines that are at least 5miles/8 km in length with a minimum of three stations and the railway provides a recognisable transport function, albeit almost exclusively for tourists and enthusiasts.

The Dartmouth Steam Railway and Riverboat Company is the best example of a fully commercial heritage railway in the UK, whilst the self-contained Keighley & Worth Valley Railway is both a heritage railway and an out-and-out preserved railway. The Swindon & Cricklade Railway and The Battlefield Line Railway remain as archetypal short preserved railways, so typical of the 1960s and 1970s whilst the Kent & East Sussex Railway transformed in April 2000 from a preserved to heritage railway with the extension of the line to Bodiam. Many tourists and day-trippers now use the railway as an entertaining and attractive mode of transport to travel from Tenterden in order to visit the National Trust Castle.

In the Americas

In the United States, heritage railways are known variously as tourist, historic, or scenic railroads. Most are remnants of original railroads, such as the Cumbres & Toltec Scenic Railroad (Colorado & New Mexico) or the Heber Valley Railroad (Utah). Others are reconstructed railroads, having been scrapped at one point and then rebuilt with tourism in mind, like the Sumpter Valley Railroad (Oregon) and the Wiscasset, Waterville & Farmington (Maine), or can be entire railroads, preserved in their original state using original structures, track, and motive power, like the Nevada Northern Railway (Nevada).

Some do not fit in the above categories, like Rio Grande Scenic Railroad, which is a sub-operation of the San Luis & Rio Grande Railroad. The SL&RG is primarily a freight operation, on former Denver & Rio Grande Western Railroad track, but owns and operates a steam locomotive and a fleet of passenger cars, most of which are painted in D&RGW colors.

Many heritage railways in the United States also function as living history museums, hosting annual reenactments of historic activities. In addition, they may feature an archive or library of railroad-related materials.

See also

- Heritage railways in Britain
- Heritage streetcar
- List of heritage railways
- Mountain railways of India
- Restored trains

External links

- UK Heritage Railways [1]
- International Working Steam [2]
- Scenic railways in France [3] mainline and tourist routes
- UK Heritage Railway Photographs [4]
- National Preservation UK's leading heritage railways forum [5]
- Hungarian Interactive Railway Museum, Budapest [6]

References

[1] http://www.heritagerailways.com/
[2] http://www.steam.dial.pipex.com/internat.htm
[3] http://about-france.com/scenic-railways.htm
[4] http://www.heritagerail.co.uk/
[5] http://railways.national-preservation.com/
[6] http://www.mavnosztalgia.hu/en/

Branch line

A **branch line** is a secondary railway line which branches off a more important through route, usually a main line. A very short branch line may be called a *spur line*. David Blyth Hanna, the first president of the Canadian National Railway, said that although most branch lines cannot pay for themselves, they are essential to make main lines pay.[1] [2]

United Kingdom

Many British branch lines were closed as a result of the "Beeching Axe" in the 1960s, although some have been re-opened as heritage railways.

The smallest branch line that is still in operation in the UK is the Stourbridge Town Branch Line from Stourbridge Junction going to Stourbridge Town. It has only one track. The journey is a third of a mile (about half of a kilometer) and the train takes around 55 s to complete its journey.

The '0 kilometre peg' marks the start of a branch line in Western Australia.

Hong Kong

Examples of spur lines in Hong Kong:

- Lok Ma Chau Spur Line
- South Tseung Kwan O Spur Line
- Racecourse Spur Line

North America

In North America, little used branch lines are often spun off from larger railroads to become new common carrier short-line railroads of their own. Throughout the United States and Canada, branch lines serve to link smaller towns or cities located too distant from the main line to be served efficiently. They were typically built to lower standards, utilizing lighter rail and shallow roadbeds when compared to main lines. In the United States, abandonment of unproductive branch lines was a byproduct of deregulation of the rail industry through the Staggers Act.

New Zealand

New Zealand once had a very extensive network of branch lines, especially in the South Island regions of Canterbury, Otago, and Southland. Many were built in the late 19th century to open up regions inland from coastal harbours and cities for farming and other economic activities. The branches in the aforementioned South Island regions were often general-purpose lines that carried predominantly agricultural traffic, but lines elsewhere were often built to serve a specific resource: on the West Coast, an extensive network of branch lines was built in rugged terrain to serve coal mines, while in the central North Island and the Bay of Plenty, lines were built inland to provide rail access to large logging operations.

Today, many of the branch lines have been closed, including almost all of the general-purpose country lines. Those that remain serve ports or industries not located near main lines such as coal mines, logging operations, large dairying factories, and steelworks. In Wellington, two branch lines exist solely for commuter passenger trains. For more, see the list of New Zealand railway lines.

Japan

There are some branch lines in Japan. The longest branch line is the 18.0 km long Saikyō Line which is a common name of the Tōhoku Main Line branch line between Akabane Station and Ōmiya Station via Musashi-Urawa Station.

Akabane Station and Ōmiya Station are also connected by the Tohoku Main Line via Urawa Station. Such branch lines are alternate routes of main lines and are connected to the main line on both ends, such as Gotemba Line (alternate route of the Tokaido Main Line), Ako, Kure, Gantoku and Ube Lines (alternate routes of the Sanyo Main Line).

References

[1] Hanna, David Blyth, Macmillan 1924

[2] Dow, Andrew, *Dow's Dictionary of Railway Quotations*, JHU Press 2006

Seymour Railway Heritage Centre

The **Seymour Railway Heritage Centre** (SRHC) is a railway preservation group based in Seymour, Victoria, Australia. The volunteer non-profit incorporated association[1] was established in 1983 to restore and preserve locomotives and rolling stock as used on the railways of Victoria.

The group is an accredited railway operator under the Victorian Rail Safety Act 2006, permitting it to move trains within its own depot.[2] The group is also accredited to maintain and provide rolling stock on the Victorian railway network,[3] running

Locomotives B74 and S303 with the 1937 *Spirit of Progress* consist, as restored by the Seymour Railway Heritage Centre in 2007

charters, tourist and railfan specials across the state with their fleet of restored trains. It is also a provider of locomotives to freight operator, El Zorro.[3]

Fleet

The Seymour Railway Heritage Centre is the custodian of a number of heritage pieces of rolling stock owned by the Victorian Government (by either VicTrack or V/Line),[4] as well as other rolling stock owned outright.

Locomotives in the custody of the group include steam locomotive J515, mainline diesels C501, S303 and B74, a number of smaller branchline T class locomotives,[3] (T320, T333, T341, T357 & T378) and smaller shunting locos Y133 and F202. Also 600hp diesel railcar DRC43.

Locomotives T357 and T320 on a Seymour Railway Heritage Centre tour

Locomotives owned by the group are: GM36, T378 & J512.

J515 & Y133 are currently on long term loan to the Victorian Goldfields Railway based in Maldon, Victoria. Whilst T333 is owned by the Victorian Goldfields Railway and on long term loan to the SRHC. T341 is also on long term loan from the Yarra Valley Tourist Railway.

Carriages in the group's collection include 1906 'E' type wooden sitting and sleeping carriages as used on the *Adelaide Express* and *The Overland*,[5] the majority of the original 1937 *Spirit of Progress* consist, and carriages used on Victorian Railways Royal Trains.

In 2007 the Seymour Railway Heritage Centre was provided with funding from the VicTrack Heritage Program for the restoration of the *Spirit of Progress* consist and heritage diesel locomotives B74 and S303,[6] and on November 25, 2007 a commemorative run was made for the 70th anniversary of the first *Spirit of Progress* service.[7]

New additions

In late August 2009, locomotives S310 and T382 were transferred to Seymour and became part of the SRHC collection. Whilst neither unit is in operational condition, as time and finances permit both units will be restored to traffic.

In early January 2010, X31 was transferred to Seymour and became part of the SRHC collection. This event has made X31 the first of its class to enter preservation. 31 will undergo a full bodywork overhaul and repaint into Victorian Railways blue and gold livery, then enter traffic as a heritage unit, believed to be used exclusively on heritage tours. Until the transfer to Seymour, X31 was part of the Pacific National fleet and had spent many months in storage.

In early March 2011, X31 returned to service as a heritage unit in Victorian Railways blue and gold livery, on hire to El Zorro.

See also

- Seymour railway station
- Tourist and Heritage Railways Act

References

[1] Seymour Railway Heritage Centre website (http://www.srhc.org.au/about.php)

[2] Public Transport Safety Victoria: List of accredited rail operators (http://www.ptsv.vic.gov.au/web26/home.nsf/AllDocs/
1B5422AA57A9E623CA2573A200017DF3?OpenDocument)

[3] V/Line: Network Service Plan - Addenda (5 December 2007) (NA_NSP_03 – R40) (http://www.vline.com.au/pdf/rna/addenda.pdf)

[4] ASSOCIATION OF TOURIST RAILWAYS: Presentation of Heritage Issues by Mr M. Ryan of the Department of Infrastructure - 29 and 30
MAY 2004 (http://home.vicnet.net.au/~atr/specials/heritage_seminar.htm)

[5] Comrails: Wooden Vestibule V&SAR Joint Stock (http://www.comrails.com/sar_carriages/a0202_se.html)

[6] MINISTER FOR PUBLIC TRANSPORT: 'SPIRIT OF PROGRESS BACK ON TRACK FOR 70TH ANNIVERSARY' - November 25, 2007
(http://www.dpc.vic.gov.au/domino/Web_Notes/newmedia.nsf/798c8b072d117a01ca256c8c0019bb01/
02d68aa2dff7cc42ca25739e00795848!OpenDocument)

[7] Seymour Railway Heritage Centre: The Official 70th Anniversary of the first run of the Spirit of Progress Tour (http://www.srhc.org.au/
tours.php?action=display&id=26)

External links

- http://www.srhc.org.au (http://www.srhc.org.au) - Official Site

South Gippsland Railway

South Gippsland Railway	
Line details	
Completed	1891
Closed	14 December 1994
Reopened	15 December 1994
Stations	Leongatha to Nyora
Rail transport in Victoria	
Closed railways	

The **South Gippsland Railway** is a tourist railway located in south Gippsland, Victoria, Australia. It controls a section of the former South Gippsland line between Nyora and Leongatha, operating services from Leongatha to Nyora via Korumburra taking around 65 minutes, trains operate on sundays, public holidays (except Good Friday and Christmas Day) and wednesdays during Victorian school holidays. The line passes through the rolling Strzelecki Ranges and lush dairy farmland.

History

The South Gippsland line was opened from Dandenong to Cranbourne in 1888 and extended to Koo Wee Rup, Nyora and Loch in 1890, Korumburra and Leongatha in 1891. The section from Lang Lang to Leongatha was transferred to the South Gippsland Railway in 1994. Freight trains continued to use the line from Dandenong as far as Koala Siding near Nyora until 1998.[1]

Projects

Station works are in progress at both Nyora and Korumburra stations. Korumburra works include the establishment of works sheds and locomotive and rollingstock stabling sheds. The work commenced in February 2009, and is scheduled to be completed by 30 June.[2]

Redhen railcar and goods rollingstock at Leongatha station

Nyora works include the repairs to the station building, including repair of internal and external wall cladding, establishment of a new Safeworking Office and Public Space / meeting facilities in the main room of the station.

Rollingstock refurbishment projects include two ABU class corridor compartment carriages being transferred to the railway,[2] and cleaning up and repainting to original colours of various goods wagons.

15 June 2010 saw the arrival of railmotor 61RM. This unit is owned by Victorian Goldfields Railway and is on a long term loan to the South Gippsland Railway for use on its services.

The South Gippsland Tourist Railway at Korrumburra loaned out K 190 during the summer of 1995-1996, repainted the engine and tender a more noticeable green. She was once again returned for service with the S.G.T.R during the summer of 1996-1997.

Future projects also include possibly restoring/reserving the line north of Nyora to Cranbourne, which is the current suburban limit for Metro trains. The line between Cranbourne and Nyora is unused and in a unusable state for trains to operate.

Regular train services

Regular weekend train services are provided by three train types. These are:

- VR Y class locomotive 135, which can haul up to two or three passenger carriages and freight rolling stock.
- VR Railmotor RM55.
- SAR Rail Motor "Red Hen" as 1 or two car sets.
- SGR also own a VR T Class diesel locomotive number 342. It is currently leased out to the Australian freight operator El Zorro.

The services are operated on Sundays, and during school holiday periods, on Sundays AND Wednesdays.

Special trains and charters

South Gippsland Railways operate regular Sunday trains on the complete Lenogatha to Nyora section. In addition, a number of "Special" rains are operated by the railway.

These include the Dinner Train, a service providing in-train hospitality, beverages and snacks, with a barbecue-style dinner evening at Korumburra Station. These services run approximately every two months.

Another special service is the Murder Mystery Train - by group booking only - providing an entertainer / host and passengers participating in character costume.

Stations

See also

- Tourist and Heritage Railways Act

References

[1] "Farewell — The Sand Train". *Newsrail* (Australian Railway Historical Society (Victorian Division)): pages 71–76. February 1998.
[2] "South Gippsland Tourist Railway: Projects" (http://www.sgr.org.au/projects. html). www.sgr.org.au. . Retrieved 2010-07-16.

External links

- South Gippsland Railway (http://www.sgr.org.au/) - (official site)

Carriage 19BE at Korumburra

Van 17CW at Korumburra

Castlemaine railway station

Main V/Line platforms at Castlemaine looking towards Melbourne

Station information	
Code	CME
Distance from Southern Cross	127.28 km (79 mi)
Operator	V/Line
Lines	Echuca Swan Hill
# Platforms	3
# Tracks	3
Status	Staffed Station
Viclink [1] profile	Link [2]

Castlemaine is a railway station in Castlemaine on the Bendigo line in Victoria, Australia. The station is located on Kennedy Street.

Platforms and services

In the morning, trains to Melbourne depart from Platform 1, while Bendigo trains depart from Platform 2, this reverses in the afternoon.

Platforms 1 and 2:

- Bendigo line - V/Line services to Southern Cross
- Bendigo line - V/Line services to Bendigo (1 service per day extends to Echuca, 2 services per day extend to Swan Hill)

Platform 3: The third track/platform is used by the Victorian Goldfields Railway, a tourist operation between Castlemaine and Maldon.

Gallery

Platform 1 and main station building, seen from Platform 2

Melbourne end of the station, Maryborough line (used for Maldon services) to the left

Victorian Goldfields Railway train on Platform 3, departing for Maldon, 'Castlemaine A' signal box at right

External links

- Victorian Station Histories - Castlemaine and Maldon Junction [3]

Elphinstone and **Taradale** stations are currently not in use and are located between Castlemaine and Malmsbury stations. **Chewton** station has been removed and was also located between Castlemaine and Malmsbury stations

Harcourt and **Ravenswood** stations are currently not in use and are located between Castlemaine and Kangaroo Flat stations.

References

[1] http://www.viclink.com.au/
[2] http://www.viclink.com.au/stop/view/20301
[3] http://www.vrhistory.com/Locations/M078-Castlemaine.pdf

Article Sources and Contributors

Maldon railway station *Source*: http://en.wikipedia.org/w/index.php?title=Maldon_railway_station *Contributors*: Billingd, Crusoe8181, D6, Dan027, Docu, Jwoodger, Peter Shearan, Rebecca, Reinthal, Tabletop, Wombat321, Wongm, Wwoods, Zzrbiker, 2 anonymous edits

Victorian Goldfields Railway *Source*: http://en.wikipedia.org/w/index.php?title=Victorian_Goldfields_Railway *Contributors*: ABCD, AaronRichard, Beetle120, Crusoe8181, Darkwind, Forenti, Jevansen, K160, Longhair, Neilc, Otownfla, Peter Horn, Plasma east, Rebecca, Rjwilmsi, Sameboat, SetiHitchHiker, Supt. of Printing, Transaus, Witchwooder, Wongm, Zzrbiker, 20 anonymous edits

Bendigo railway line *Source*: http://en.wikipedia.org/w/index.php?title=Bendigo_railway_line *Contributors*: 712M, Alex1991, Amalas, AxSkov, Axpde, Betacommand, Bryan Derksen, Crusoe8181, Dan027, Garyvines, Longhair, Merveism, MoondyneAWB, Peter Horn, Rebecca, Reinthal, Rick Doodle, Rom rulz424, Samantha of Cardyke, Sameboat, Scottius11, Sheepunderscore, Signalhead, Somebody in the WWW, Supt. of Printing, Wongm, Woohookitty, 12 anonymous edits

Castlemaine, Victoria *Source*: http://en.wikipedia.org/w/index.php?title=Castlemaine%2C_Victoria *Contributors*: 3mta3, Afterwriting, Aka4761, Alex43223, Angela, Biatch, Bidgee, Bluerasberry, Brucepython, Ca47, Cam Walker, Cassowary, Clappingsimon, Clarkk, Crusoe8181, Deflective, Epistemos, EyeSerene, Frankenstien, Gatoclass, Gervo1865, Grahamec, Greenshed, Halidayn, Helpful Dave, Hmains, Jevansen, Jonahallen, Josh Parris, Kwamikagami, Leujohn, Longhair, Marco79, Matilda, Mattinbgn, Melburnian, Merovingian, MichaelBillington, Monkeymox, MulgaBill, Neutrality, Newsbingo, Orderinchaos 2, Packardstown, Patrick.underwood, Phobdog, Piano non troppo, Polypipe Wrangler, RedWolf, Rich Farmbrough, RichardTait, Robert Merkel, Rocketfrog, Roke, Rulesfan, RyleyB22, Scottius11, Smithy07, Sting au, TPK, Tassedethe, That Guy, From That Show!, The wub, Uncle Milty, Xcentaur, Zzrbiker, 86 anonymous edits

Muckleford railway station *Source*: http://en.wikipedia.org/w/index.php?title=Muckleford_railway_station *Contributors*: Benjamus, Billingd, Crusoe8181, D6, Dan027, Wongm, Zzrbiker

Rail transport *Source*: http://en.wikipedia.org/w/index.php?title=Rail_transport *Contributors*: *drew, -Majestic-, 1122334455, 12.234.49.xxx, 16@r, A bit iffy, Aaaallleex, Abby, Acela2038, Acprail, Acroterion, AdamW, Adashiel, Ae-a, Aesopos, Ahoerstemeier, Ahuskay, Aitias, Alansohn, Alco83, Aleator, AlexPlante, Alexdan loghin, Alextrevelian 006, Allreet, AllyUnion, Altenmann, Alucard (Dr.), Amakuru, AmyzzXX, Andreas Kaganov, Andrewlim1, Andrewpmk, Andy Dingley, Anlace, Anne McDermott, Antandrus, Anthony Appleyard, Armando, ArnoldReinhold, Arsenikk, Arthur Ellis, Aseemsjohri, Ashishbijwe24, Aude, Autonerd, AxelBoldt, Ayushrocks6, BD2412, BRG, Backifran, Bambuway, Basketball1776, Bastin, Bathrobe, Bebestbe, Beland, Berendale2, Bermicourt, Besselfunctions, Beyond silence, Bhadani, Bhtpbank, Bigdumbdinosaur, Biggus wickus, Biker Biker, Bikramshergill, Bill37212, Birdhurst, Biscuittin, Bkell, Black Kite, Blehfu, Blimpguy, Bluerasberry, Bobo192, Bogdangiusca, Borgx, Brambo, Brhaspati, Brian Pearson, Brian evans, BruceDLimber, Bryan Derksen, Btornado, Bulleid Pacific, Burgher, Camboxer, Can't sleep, clown will eat me, CanadianLinuxUser, Canterbury, Casey56, Caseyjonz, Cazort, Cecropia, Celardore, Central Data Bank, Cessator, Chaleyer61, ChanceButler, Charlottesometimes, Cheekdog, Chevin, Chris55, Christian List, Chun-hian, Civil Engineer III, Claude girardin, Cluth, Cmapm, Cmichael, Colonies Chris, Complete fanatic, Conversion script, Coolhawks88, Coosbane, Corpx, Corti, Courcelles, CrossHouses, CryptoDerk, Cwolfsheep, Cynical, DAJF, Damirgraffiti, Dancter, Daniel C, Dark Shikari, DarkFireTaker, Das48, Daveswagon, David.Monniaux, DavidArthur, Ddstretch, Dealerofsalvation, Dearmstrong, Deb, Dellarb, Delldot, DennisJOBrien@yahoo.com, Dethme0w, Devin.phillips, DianeSelvy, DigbyDalton, Dnalrom123, Dnfenner, Doc glasgow, Download, Dp67, Dragomiloff, Dratman, Dsteckelberg, Duncharris, Dysprosia, ESkog, EarthFurst, EdJogg, Edgar181, El C, ElectraFlarefire, Emdx, Emperor Genius, Emsox, Engi08, Epbr123, Epolk, Equinox137, Eric-Wester, Exe, Expertz123, Ezhiukas 7, FT2, Faradayplank, Farjiadmi, Ferrierd, Feydey, Finavon, Finngall, Fixerofnatasha, FlavrSavr, Flux.books, Foodman, Fortdj33, Frankenpuppy, Fred Bauder, Fredericsalve, Frixa, Fudoreaper, Furrykef, Fyrael, G-Man, Gandydancer, Gary D, Gbtojun, Gdo01, Geoff NoNick, Gerfriedc, Geschichte, Globalsolidarity, Golbez, Grafen, Graham87, Grim23, Gsaup, Gtstricky, Gun Powder Ma, Guybrarian, Gwernol, Gökhan, Hairy Dude, Harjasusi, Haroonn1, Harpblaster, Harperbruce, Hclincha, HenryLi, Hephaestos, Hmrox, Hoo man, Hu, Hugh Manatee, Hulagutten, Hydriotaphia, Ianp1950, Ichol, Ishlan, J.delanoy, JCam, JNW, JRR Trollkien, JYolkowski, JackLumber, Jacob.jose, Jallason, Jameswilson, Japman6, Jarle fagerheim, Jax184, Jclemens, Jcuk, Jdcook, Jdforrester, Jeepday, Jeffakolb, Jeremy5561, Jevon, Jhonchris1981, Jim, Jimp, Jinty182, Jmorgan, JoeK, Joethespaz13, Joey Roe, John, JohnOwens, Johnasher, Jojit fb, Jonathan.s.kt, JorgeGG, Joshua, Jovianeye, Jpgordon, Jpk, Jrleighton, Jstreutker, Juggleandhope, JuliusAugustus, Jusjih, Juxo, KF, KVDP, Kahuzi, Kaibabsquirrel, Kalemika, Kalixiri, Katsp8, Kazvorpal, Kdar, Keeper76, Keidansky, Keithmall, Kelisi, Kelly Martin, Keotaman, Kingpin13, Kkmd, Krath, Krigjsman, Ksyrie, Kumioko, Kungfuadam, Kuru, L blue l, La Pianista, Lafleure355, Lars T., LeaveSleaves, Leefletcher, LeoO3, Leonard G., Leptictidium, Levineps, Lightmouse, Lindahq, Lockesdonkey, Longhair, Lord Bob, Loren36, LorenzoB, Lost on belmont, Lost tourist, Lot29, Lovemehateme123, Luna Santin, MNboy, MPD01605, Madcoverboy, Madhero88, MakeChooChooGoNow, Mamawrites, Man vyi, Mangoe, Manop, Maranjc, MarsRover, Marshman, Matt von Furrie, Mattisse, MauriceJFox3, Maury Markowitz, Mav, Mayooranathan, Mdd4696, Megaman en m, Mervyn, Mgaved, Michael Hardy, MichaelBillington, Michaelduly, Mike Dill, Mikeasolomon, Mindspillage, Minesweeper, Mitul0520, MoO, Moebiusuibeom-en, Moncrief, Monedula, Morven, Moyogo, Murray Langton, N2e, NE2, Naddy, NameThatWorks, Nanana123456789, Nat Krause, Nerfer, NicholasJones, Nick Number, Nicolae Coman, Nk, Noexit, Noisy, Northumbrian, Nricardo, Nunh-huh, Oanabay04, Oceanhahn, Olahus, OlavN, Olivier, Omnedon, Ortolan88, PFHLai, PNRStreamliner, PReinie, Pak bahadur, Patoune08, Patrick, Paul Matthews, Paul W, PaulDrye, Perdelsky, Peter Horn, Peterkingiron, Pgk, Phlebas, Piano non troppo, Pickle UK, Pietrow, Pilgaard, Pilotguy, Pinethicket, Pion, Piotrus, Playclever, Plutochaun, Pne, Pointer1, Pressforaction, ProhibitOnions, Prolog, Psopmaz, Psublue23, Qertis, QuantumEngineer, Quintote, R'n'B, RFBailey, RTucker, Rafikk, Randall uob, Regregex, Rememberway, RevRagnarok, Reward, Rich Farmbrough, Rich4560, Rickyrab, Rigadoun, Rjensen, Rjmccall, Rjstott, Rjwilmsi, Rmhermen, RobHarding, Robert Merkel, Rods42, Rojomoke, Roke, Rrburke, Rror, Rsrikanth05, Runewiki777, S.K., SEWilco, SNIyer12, SPUI, ST47, Salamurai, Sam Hocevar, Sameer rana100, Samuel 69105, Samuell, Sandman30s, Sarah Waggoner, Sarah tonen, Sc147, ScAvenger lv, Scoot-Overload, Scoutersig, Seaphoto, SebbeSwe, Sentorment, Sergio.solar, Sgfaig, Sharadbob, SheepNotGoats, Sheldwich14, Signalhead, SilkTork, Silverjonny, SimonP, Skapur, Slambo, Smalljim, Smerlinare, Smurrayinchester, Snottily, Snowmanradio, Soliloquial, Sonett72, Sp33dyphil, Spagf5, Spangineer, Spearhead, Spellchecker, SqueakBox, SriMesh, Srtig, Starbois, Steam-loco, Stefan2, Steinberger, Stephen C. Carlson, Stragermont, Struway2, Stumpytrain, Sulzer55, Supt. of Printing, Suri 100, Svenman, Svetovid, SwalleyD, Syd1435, Symane, TAS, THEN WHO WAS PHONE?, Tabletop, Tagishsimon, TaintedMustard, Takamaxa, Taksim25, Tangent747, Taoster, Tarquin, Tassedethe, Tatakubuntu, Tcr25, Tedernst, Tellyaddict, Template namespace initialisation script, TerryElliott, Textorus, Tgv8925, The Arbiter, The Wild Falcon, The undertow, TheBourtreehillian, TheEgyptian, TheListUpdater, Theda, Thephotoplayer, ThomasAndrewNimmo, Tide rolls, Tim PF, Timc, Tmorgan147, Toiyabe, Tony1, TonyTheTiger, Train2104, Trekphiler, Ttsalo, Tudorminstrel, Ulric1313, Undead1, Vapmachado, Vcwizard, Vegas949, Villager57, Voyager, WOSlinker, Wahiba, Wavelength, Wdfarmer, Webnetprof, Wereon, WiJG?, WikHead, Wiki Raja, Wikipelli, William Grimes, Wolfkeeper, Wongm, Woodenships, Woohookitty, Wspencer11, Wuhwuzdat, Yelyos, Yossarian, Yvwv, Zhou Yu, Zoney, Zzuuzz, ŠJů, Јованвћ, 807 anonymous edits

Heritage railway *Source*: http://en.wikipedia.org/w/index.php?title=Heritage_railway *Contributors*: 7severn7, Ahsaninam, Aldis90, Arpingstone, Arsenikk, Bjf, Catgut, Chaser, Chzz, Ciudad jardin, CommonsDelinker, Coyau, D&RGW 223, Dace83, Duncharris, Fila934, Foxmarks, Fungusflame, G-Man, Georgeccampbell, Gwernol, Harveyqs, Hugh Nightingale, Iridescent, Jack Rance, JeffyJeffyMan2004, JeremyA, Jimfbleak, Jp347, Kneiphof, Lancevortex, Lightmouse, Lucasm, Magnus Manske, Michael Hardy, Michael Johnson, MickMacNee, Montrealais, Mulad, Myrtone86, N2e, Natpres, Nevilley, Ngc1976, Nick Number, Noirish, Numbo3, Optimist on the run, Our Phellap, Pencefn, Peter Horn, Pv, Reedy, Renata, Rjd0060, Ronstar308, Ryan Roos, Sandolsky, Schebesch, Sebwite, Sfnhltb, Simply south, Smjg, Srwalden, Supt. of Printing, Telsa, The Tom, Thryduulf, Vantey, Welsh, Wittyname, Wongm, Woohookitty, Zundark, 60 anonymous edits

Branch line *Source*: http://en.wikipedia.org/w/index.php?title=Branch_line *Contributors*: Alanmak, Alpha Ralpha Boulevard, Axver, BesigedB, DAJF, FrankieG123, Jp347, KJRehberg, Loganberry, Magioladitis, MickMacNee, Nachoman-au, Nictrain, Optimist on the run, Penpen, Peter Horn, Pigsonthewing, Pne, Quatro Valvole, Robofish, Shawn in Montreal, SiobhanHansa, Slambo, SouthernElectric, Surge79uwf, Sweetstudent, TigerShark, ŠJů, , 7 anonymous edits

Seymour Railway Heritage Centre *Source*: http://en.wikipedia.org/w/index.php?title=Seymour_Railway_Heritage_Centre *Contributors*: Billingd, Cpdbear, Crusoe8181, Mattinbgn, Mild Bill Hiccup, Plasma east, Rjwilmsi, Srhc01, Transaus, WikiSandbox1, Wongm, 10 anonymous edits

South Gippsland Railway *Source*: http://en.wikipedia.org/w/index.php?title=South_Gippsland_Railway *Contributors*: Axpde, Billingd, Crusoe8181, Dan027, Forthevline, Grahamec, Peter Horn, Plasma east, Robert Ashworth, Sameboat, Supt. of Printing, Transaus, Wongm, Zzrbiker, 20 anonymous edits

Castlemaine railway station *Source*: http://en.wikipedia.org/w/index.php?title=Castlemaine_railway_station *Contributors*: 712M, Alex1991, Amalas, Blanchardb, Crusoe8181, Dan027, Deor, Forthevline, Grahamec, Grey Shadow, Longhair, Maias, RedWolf, Rom rulz424, Scottius11, Somebody in the WWW, Supt. of Printing, Tanarrifujitsu, TheParanoidOne, Wongm, Zzrbiker, 12 anonymous edits

Image Sources, Licenses and Contributors

image:VictorianGoldfieldsRailwaylogo.png *Source*: http://en.wikipedia.org/w/index.php?title=File:VictorianGoldfieldsRailwaylogo.png *License*: unknown *Contributors*: User:Zzrbiker

image:maldon-railway-station.jpg *Source*: http://en.wikipedia.org/w/index.php?title=File:Maldon-railway-station.jpg *License*: Attribution *Contributors*: Dan027

Image:VictorianGoldfieldsRailwaylogo.png *Source*: http://en.wikipedia.org/w/index.php?title=File:VictorianGoldfieldsRailwaylogo.png *License*: unknown *Contributors*: User:Zzrbiker

Image:J515atMaldon.jpg *Source*: http://en.wikipedia.org/w/index.php?title=File:J515atMaldon.jpg *License*: GNU Free Documentation License *Contributors*: Endo-Plexor, Zzrbiker

Image:VGR reopening to Castlemaine.jpg *Source*: http://en.wikipedia.org/w/index.php?title=File:VGR_reopening_to_Castlemaine.jpg *License*: GNU Free Documentation License *Contributors*: Zzrbiker

Image:K160 at Castlemaine.jpg *Source*: http://en.wikipedia.org/w/index.php?title=File:K160_at_Castlemaine.jpg *License*: Public Domain *Contributors*: Original uploader was Zzrbiker at en.wikipedia

Image:Vrmap47vgr.png *Source*: http://en.wikipedia.org/w/index.php?title=File:Vrmap47vgr.png *License*: Public Domain *Contributors*: Zzrbiker

Image:J515 cab window.jpg *Source*: http://en.wikipedia.org/w/index.php?title=File:J515_cab_window.jpg *License*: GNU Free Documentation License *Contributors*: User:Zzrbiker

Image:MaldonRailwayStation.jpg *Source*: http://en.wikipedia.org/w/index.php?title=File:MaldonRailwayStation.jpg *License*: Creative Commons Attribution-Sharealike 2.5 *Contributors*: User:SetiHitchHiker

Image:Vicrailmap-bendigo.png *Source*: http://en.wikipedia.org/w/index.php?title=File:Vicrailmap-bendigo.png *License*: GNU Free Documentation License *Contributors*: User:Wongm

File:Big-Hill-tunnel-Bendigo-line.jpg *Source*: http://en.wikipedia.org/w/index.php?title=File:Big-Hill-tunnel-Bendigo-line.jpg *License*: GNU Free Documentation License *Contributors*: User:Wongm

File:Taradale7.jpg *Source*: http://en.wikipedia.org/w/index.php?title=File:Taradale7.jpg *License*: unknown *Contributors*: Alex1991

File:Kyneton1.jpg *Source*: http://en.wikipedia.org/w/index.php?title=File:Kyneton1.jpg *License*: unknown *Contributors*: Alex1991

File:Kyneton11.jpg *Source*: http://en.wikipedia.org/w/index.php?title=File:Kyneton11.jpg *License*: unknown *Contributors*: Alex1991

File:Castlemaine-station-Melbourne-end.jpg *Source*: http://en.wikipedia.org/w/index.php?title=File:Castlemaine-station-Melbourne-end.jpg *License*: GNU Free Documentation License *Contributors*: User:Wongm

File:Castlemaine from burke and wills memorial lookout.jpg *Source*: http://en.wikipedia.org/w/index.php?title=File:Castlemaine_from_burke_and_wills_memorial_lookout.jpg *License*: Creative Commons Attribution-Sharealike 3.0 *Contributors*: User:Biatch

File:Red pog.svg *Source*: http://en.wikipedia.org/w/index.php?title=File:Red_pog.svg *License*: Public Domain *Contributors*: User:Andux

File:Loudspeaker.svg *Source*: http://en.wikipedia.org/w/index.php?title=File:Loudspeaker.svg *License*: Public Domain *Contributors*: Bayo, Gmaxwell, Husky, Iamunknown, Myself488, Nethac DIU, Omegatron, Rocket000, The Evil IP address, Wouterhagens, 9 anonymous edits

File:Barker street castlemaine 1908.jpg *Source*: http://en.wikipedia.org/w/index.php?title=File:Barker_street_castlemaine_1908.jpg *License*: Public Domain *Contributors*: Biatch

File:CastlemaineTownHall.JPG *Source*: http://en.wikipedia.org/w/index.php?title=File:CastlemaineTownHall.JPG *License*: Creative Commons Attribution-Sharealike 3.0 *Contributors*: User:Mattinbgn

Image:CastlemainePostOffice.JPG *Source*: http://en.wikipedia.org/w/index.php?title=File:CastlemainePostOffice.JPG *License*: Creative Commons Attribution-Sharealike 3.0 *Contributors*: User:Mattinbgn

File:CastlemaineArtGallery.JPG *Source*: http://en.wikipedia.org/w/index.php?title=File:CastlemaineArtGallery.JPG *License*: Creative Commons Attribution-Sharealike 3.0 *Contributors*: User:Mattinbgn

File:CastlemaineStation.JPG *Source*: http://en.wikipedia.org/w/index.php?title=File:CastlemaineStation.JPG *License*: Public Domain *Contributors*: Gareth

File:CastlemaineHarryLawsonBust.JPG *Source*: http://en.wikipedia.org/w/index.php?title=File:CastlemaineHarryLawsonBust.JPG *License*: Creative Commons Attribution-Sharealike 3.0 *Contributors*: User:Mattinbgn

File:BNSF 5350 20040808 Prairie du Chien WI.jpg *Source*: http://en.wikipedia.org/w/index.php?title=File:BNSF_5350_20040808_Prairie_du_Chien_WI.jpg *License*: Creative Commons Attribution-Sharealike 2.0 *Contributors*: User:Slambo

File:DeutscheBahn gobeirne.jpg *Source*: http://en.wikipedia.org/w/index.php?title=File:DeutscheBahn_gobeirne.jpg *License*: Creative Commons Attribution 2.5 *Contributors*: User:gobeirne

Image:Three rail tracks 350.jpg *Source*: http://en.wikipedia.org/w/index.php?title=File:Three_rail_tracks_350.jpg *License*: Public Domain *Contributors*: Original uploader was w:en:User:G-ManG-Man at en.wikipedia Later version(s) were uploaded by w:en:User:ST47ST47 at en.wikipedia.

File:Brno, Brno Město, historická koňská tramvaj.jpg *Source*: http://en.wikipedia.org/w/index.php?title=File:Brno,_Brno_Město,_historická_koňská_tramvaj.jpg *License*: Creative Commons Attribution-Sharealike 2.5 *Contributors*: User:Aktron/Nápověda

File:5051 Earl Bathurst Cocklewood Harbour.jpg *Source*: http://en.wikipedia.org/w/index.php?title=File:5051_Earl_Bathurst_Cocklewood_Harbour.jpg *License*: unknown *Contributors*: Alex at kms, Duncharris, Geof Sheppard, Nilfanion, RHaworth, Railwayfan2005, Shortfatlad, 1 anonymous edits

File:CTA tracks.jpg *Source*: http://en.wikipedia.org/w/index.php?title=File:CTA_tracks.jpg *License*: Creative Commons Attribution-Sharealike 2.5 *Contributors*: User:Dschwen

File:Ireland - Dublin - Tram.jpg *Source*: http://en.wikipedia.org/w/index.php?title=File:Ireland_-_Dublin_-_Tram.jpg *License*: Creative Commons Attribution 2.0 *Contributors*: User:CGPGrey

File:2TE10U Russian Locomotive.jpg *Source*: http://en.wikipedia.org/w/index.php?title=File:2TE10U_Russian_Locomotive.jpg *License*: Public Domain *Contributors*: User:Anthony Ivanoff

File:HŽ 7123 series DMU (06).JPG *Source*: http://en.wikipedia.org/w/index.php?title=File:HŽ_7123_series_DMU_(06).JPG *License*: Public Domain *Contributors*: User:Orlovic

File:InterCity2 - passenger car interior.jpg *Source*: http://en.wikipedia.org/w/index.php?title=File:InterCity2_-_passenger_car_interior.jpg *License*: unknown *Contributors*: Jonik at en.wikipedia

File:Wagons 550.jpg *Source*: http://en.wikipedia.org/w/index.php?title=File:Wagons_550.jpg *License*: unknown *Contributors*: Original uploader was G-Man at en.wikipedia

Image:Railroad-Gyula-b.jpg *Source*: http://en.wikipedia.org/w/index.php?title=File:Railroad-Gyula-b.jpg *License*: Creative Commons Attribution-Sharealike 2.5 *Contributors*: User:Beyond silence

Image:CTA loop junction.jpg *Source*: http://en.wikipedia.org/w/index.php?title=File:CTA_loop_junction.jpg *License*: Creative Commons Attribution-Sharealike 2.5 *Contributors*: User:Dschwen

File:Eastbound over SCB.jpg *Source*: http://en.wikipedia.org/w/index.php?title=File:Eastbound_over_SCB.jpg *License*: GNU Free Documentation License *Contributors*: Davepape, Gürbetaler, Iain Bell, Ronaldino, Thryduulf, Voyager, 2 anonymous edits

File:HBD DD1.jpg *Source*: http://en.wikipedia.org/w/index.php?title=File:HBD_DD1.jpg *License*: Creative Commons Zero *Contributors*: User:SwalleyD

File:Rail-semaphore-signal-Dave-F.jpg *Source*: http://en.wikipedia.org/w/index.php?title=File:Rail-semaphore-signal-Dave-F.jpg *License*: Creative Commons Attribution 2.0 *Contributors*: Original uploader was AmosWolfe at en.wikipedia

File:Chhatrapati Shivaji Terminus (Victoria Terminus).jpg *Source*: http://en.wikipedia.org/w/index.php?title=File:Chhatrapati_Shivaji_Terminus_(Victoria_Terminus).jpg *License*: Creative Commons Attribution-Sharealike 3.0 *Contributors*: User:Jovianeye

File:UP 6670.jpg *Source*: http://en.wikipedia.org/w/index.php?title=File:UP_6670.jpg *License*: Creative Commons Attribution-Sharealike 2.0 *Contributors*: terry cantrell

File:Train wreck at Montparnasse 1895.jpg *Source*: http://en.wikipedia.org/w/index.php?title=File:Train_wreck_at_Montparnasse_1895.jpg *License*: Public Domain *Contributors*: Studio Lévy and Sons (Studio Lévy & fils)

File:Stanhope Station Railway Lines.jpg *Source*: http://en.wikipedia.org/w/index.php?title=File:Stanhope_Station_Railway_Lines.jpg *License*: Public Domain *Contributors*: Tellyaddict

Image:heritage.rail.750pix.jpg *Source*: http://en.wikipedia.org/w/index.php?title=File:Heritage.rail.750pix.jpg *License*: Public Domain *Contributors*: Andy Dingley, Arpingstone, Chris j wood, Duncharris, Geof Sheppard, Shortfatlad, 2 anonymous edits

File:DHR 780 on Batasia Loop 05-02-21 08.jpeg *Source*: http://en.wikipedia.org/w/index.php?title=File:DHR_780_on_Batasia_Loop_05-02-21_08.jpeg *License*: Creative Commons Attribution 2.5 *Contributors*: AHEMSLTD, Denniss, Diwas, Roland zh

image:KhyberRailway 02.jpg *Source*: http://en.wikipedia.org/w/index.php?title=File:KhyberRailway_02.jpg *License*: Creative Commons Attribution-Sharealike 3.0 *Contributors*: User:MaltaGC

Image:Bluebell favourite Stepney in front of the 1960's shed - geograph.org.uk - 1007122.jpg *Source*: http://en.wikipedia.org/w/index.php?title=File:Bluebell_favourite_Stepney_in_front_of_the_1960's_shed_-_geograph.org.uk_-_1007122.jpg *License*: Attribution *Contributors*: Ashley Dace

Image:Railway kilometre peg.jpg *Source*: http://en.wikipedia.org/w/index.php?title=File:Railway_kilometre_peg.jpg *License*: Creative Commons Attribution-Sharealike 2.5 *Contributors*: User:Nachoman-au

Image:Spirit-of-progress-70th-anniversary-2007.jpg *Source*: http://en.wikipedia.org/w/index.php?title=File:Spirit-of-progress-70th-anniversary-2007.jpg *License*: GNU Free Documentation License *Contributors*: User:Wongm

Image:Restored victorian railways train.jpg *Source*: http://en.wikipedia.org/w/index.php?title=File:Restored_victorian_railways_train.jpg *License*: Creative Commons Attribution-Sharealike 2.5 *Contributors*: User:Wongm

File:South-Gippsland-Railway-Leongatha.jpg *Source*: http://en.wikipedia.org/w/index.php?title=File:South-Gippsland-Railway-Leongatha.jpg *License*: Creative Commons Attribution-Sharealike 3.0 *Contributors*: User:Wongm

File:19BE air conditioned.jpg *Source*: http://en.wikipedia.org/w/index.php?title=File:19BE_air_conditioned.jpg *License*: Creative Commons Attribution-Sharealike 3.0 *Contributors*: User:Wongm

File:17CW van.jpg *Source*: http://en.wikipedia.org/w/index.php?title=File:17CW_van.jpg *License*: Creative Commons Attribution-Sharealike 3.0 *Contributors*: User:Wongm

Image:Viclink Station Board.PNG *Source*: http://en.wikipedia.org/w/index.php?title=File:Viclink_Station_Board.PNG *License*: Public Domain *Contributors*: Lakeyboy

Image:CastlemaineStation.JPG *Source*: http://en.wikipedia.org/w/index.php?title=File:CastlemaineStation.JPG *License*: Public Domain *Contributors*: Gareth

Image:Castlemaine Platform 1.JPG *Source*: http://en.wikipedia.org/w/index.php?title=File:Castlemaine_Platform_1.JPG *License*: Creative Commons Attribution-Sharealike 3.0 *Contributors*: User:Zzrbiker

Image:Castlemaine-station-Melbourne-end.jpg *Source*: http://en.wikipedia.org/w/index.php?title=File:Castlemaine-station-Melbourne-end.jpg *License*: GNU Free Documentation License *Contributors*: User:Wongm

Image:Maldon_train_at_Castlemaine.JPG *Source*: http://en.wikipedia.org/w/index.php?title=File:Maldon_train_at_Castlemaine.JPG *License*: GNU Free Documentation License *Contributors*: Zzrbiker (talk). Original uploader was Zzrbiker at en.wikipedia

License

GNU Free Documentation License

As of July 15, 2009 Wikipedia has moved to a dual-licensing system that supersedes the previous GFDL only licensing. In short, this means that text licensed under the GFDL can no longer be imported to Wikipedia. Additionally, text contributed after that date can not be exported under the GFDL license. See Wikipedia:Licensing update for further information.

Version 1.3, 3 November 2008 Copyright (C) 2000, 2001, 2002, 2007, 2008 Free Software Foundation, Inc. <http://fsf.org/>
Everyone is permitted to copy and distribute verbatim copies of this license document, but changing it is not allowed.

0. PREAMBLE

The purpose of this License is to make a manual, textbook, or other functional and useful document "free" in the sense of freedom: to assure everyone the effective freedom to copy and redistribute it, with or without modifying it, either commercially or noncommercially. Secondarily, this License preserves for the author and publisher a way to get credit for their work, while not being considered responsible for modifications made by others.
This License is a kind of "copyleft", which means that derivative works of the document must themselves be free in the same sense. It complements the GNU General Public License, which is a copyleft license designed for free software.
We have designed this License in order to use it for manuals for free software, because free software needs free documentation: a free program should come with manuals providing the same freedoms that the software does. But this License is not limited to software manuals; it can be used for any textual work, regardless of subject matter or whether it is published as a printed book. We recommend this License principally for works whose purpose is instruction or reference.

1. APPLICABILITY AND DEFINITIONS

This License applies to any manual or other work, in any medium, that contains a notice placed by the copyright holder saying it can be distributed under the terms of this License. Such a notice grants a world-wide, royalty-free license, unlimited in duration, to use that work under the conditions stated herein. The "Document", below, refers to any such manual or work. Any member of the public is a licensee, and is addressed as "you". You accept the license if you copy, modify or distribute the work in a way requiring permission under copyright law.
A "Modified Version" of the Document means any work containing the Document or a portion of it, either copied verbatim, or with modifications and/or translated into another language.
A "Secondary Section" is a named appendix or a front-matter section of the Document that deals exclusively with the relationship of the publishers or authors of the Document to the Document's overall subject (or to related matters) and contains nothing that could fall directly within that overall subject. (Thus, if the Document is in part a textbook of mathematics, a Secondary Section may not explain any mathematics.) The relationship could be a matter of historical connection with the subject or with related matters, or of legal, commercial, philosophical, ethical or political position regarding them.
The "Invariant Sections" are certain Secondary Sections whose titles are designated, as being those of Invariant Sections, in the notice that says that the Document is released under this License. If a section does not fit the above definition of Secondary then it is not allowed to be designated as Invariant. The Document may contain zero Invariant Sections. If the Document does not identify any Invariant Sections then there are none.
The "Cover Texts" are certain short passages of text that are listed, as Front-Cover Texts or Back-Cover Texts, in the notice that says that the Document is released under this License. A Front-Cover Text may be at most 5 words, and a Back-Cover Text may be at most 25 words.
A "Transparent" copy of the Document means a machine-readable copy, represented in a format whose specification is available to the general public, that is suitable for revising the document straightforwardly with generic text editors or (for images composed of pixels) generic paint programs or (for drawings) some widely available drawing editor, and that is suitable for input to text formatters or for automatic translation to a variety of formats suitable for input to text formatters. A copy made in an otherwise Transparent file format whose markup, or absence of markup, has been arranged to thwart or discourage subsequent modification by readers is not Transparent. An image format is not Transparent if used for any substantial amount of text. A copy that is not "Transparent" is called "Opaque".
Examples of suitable formats for Transparent copies include plain ASCII without markup, Texinfo input format, LaTeX input format, SGML or XML using a publicly available DTD, and standard-conforming simple HTML, PostScript or PDF designed for human modification. Examples of transparent image formats include PNG, XCF and JPG. Opaque formats include proprietary formats that can be read and edited only by proprietary word processors, SGML or XML for which the DTD and/or processing tools are not generally available, and the machine-generated HTML, PostScript or PDF produced by some word processors for output purposes only.
The "Title Page" means, for a printed book, the title page itself, plus such following pages as are needed to hold, legibly, the material this License requires to appear in the title page. For works in formats which do not have any title page as such, "Title Page" means the text near the most prominent appearance of the work's title, preceding the beginning of the body of the text.
The "publisher" means any person or entity that distributes copies of the Document to the public.
A section "Entitled XYZ" means a named subunit of the Document whose title either is precisely XYZ or contains XYZ in parentheses following text that translates XYZ in another language. (Here XYZ stands for a specific section name mentioned below, such as "Acknowledgements", "Dedications", "Endorsements", or "History".) To "Preserve the Title" of such a section when you modify the Document means that it remains a section "Entitled XYZ" according to this definition.
The Document may include Warranty Disclaimers next to the notice which states that this License applies to the Document. These Warranty Disclaimers are considered to be included by reference in this License, but only as regards disclaiming warranties: any other implication that these Warranty Disclaimers may have is void and has no effect on the meaning of this License.

2. VERBATIM COPYING

You may copy and distribute the Document in any medium, either commercially or noncommercially, provided that this License, the copyright notices, and the license notice saying this License applies to the Document are reproduced in all copies, and that you add no other conditions whatsoever to those of this License. You may not use technical measures to obstruct or control the reading or further copying of the copies you make or distribute. However, you may accept compensation in exchange for copies. If you distribute a large enough number of copies you must also follow the conditions in section 3.
You may also lend copies, under the same conditions stated above, and you may publicly display copies.

3. COPYING IN QUANTITY

If you publish printed copies (or copies in media that commonly have printed covers) of the Document, numbering more than 100, and the Document's license notice requires Cover Texts, you must enclose the copies in covers that carry, clearly and legibly, all these Cover Texts: Front-Cover Texts on the front cover, and Back-Cover Texts on the back cover. Both covers must also clearly and legibly identify you as the publisher of these copies. The front cover must present the full title with all words of the title equally prominent and visible. You may add other material on the covers in addition. Copying with changes limited to the covers, as long as they preserve the title of the Document and satisfy these conditions, can be treated as verbatim copying in other respects.
If the required texts for either cover are too voluminous to fit legibly, you should put the first ones listed (as many as fit reasonably) on the actual cover, and continue the rest onto adjacent pages.
If you publish or distribute Opaque copies of the Document numbering more than 100, you must either include a machine-readable Transparent copy along with each Opaque copy, or state in or with each Opaque copy a computer-network location from which the general network-using public has access to download using public-standard network protocols a complete Transparent copy of the Document, free of added material. If you use the latter option, you must take reasonably prudent steps, when you begin distribution of Opaque copies in quantity, to ensure that this Transparent copy will remain thus accessible at the stated location until at least one year after the last time you distribute an Opaque copy (directly or through your agents or retailers) of that edition to the public.
It is requested, but not required, that you contact the authors of the Document well before redistributing any large number of copies, to give them a chance to provide you with an updated version of the Document.

4. MODIFICATIONS

You may copy and distribute a Modified Version of the Document under the conditions of sections 2 and 3 above, provided that you release the Modified Version under precisely this License, with the Modified Version filling the role of the Document, thus licensing distribution and modification of the Modified Version to whoever possesses a copy of it. In addition, you must do these things in the Modified Version:

A. Use in the Title Page (and on the covers, if any) a title distinct from that of the Document, and from those of previous versions (which should, if there were any, be listed in the History section of the Document). You may use the same title as a previous version if the original publisher of that version gives permission.
B. List on the Title Page, as authors, one or more persons or entities responsible for authorship of the modifications in the Modified Version, together with at least five of the principal authors of the Document (all of its principal authors, if it has fewer than five), unless they release you from this requirement.
C. State on the Title page the name of the publisher of the Modified Version, as the publisher.
D. Preserve all the copyright notices of the Document.
E. Add an appropriate copyright notice for your modifications adjacent to the other copyright notices.
F. Include, immediately after the copyright notices, a license notice giving the public permission to use the Modified Version under the terms of this License, in the form shown in the Addendum below.
G. Preserve in that license notice the full lists of Invariant Sections and required Cover Texts given in the Document's license notice.
H. Include an unaltered copy of this License.
I. Preserve the section Entitled "History", Preserve its Title, and add to it an item stating at least the title, year, new authors, and publisher of the Modified Version as given on the Title Page. If there is no section Entitled "History" in the Document, create one stating the title, year, authors, and publisher of the Document as given on its Title Page, then add an item describing the Modified Version as stated in the previous sentence.
J. Preserve the network location, if any, given in the Document for public access to a Transparent copy of the Document, and likewise the network locations given in the Document for previous versions it was based on. These may be placed in the "History" section. You may omit a network location for a work that was published at least four years before the Document itself, or if the original publisher of the version it refers to gives permission.
K. For any section Entitled "Acknowledgements" or "Dedications", Preserve the Title of the section, and preserve in the section all the substance and tone of each of the contributor acknowledgements and/or dedications given therein.
L. Preserve all the Invariant Sections of the Document, unaltered in their text and in their titles. Section numbers or the equivalent are not considered part of the section titles.
M. Delete any section Entitled "Endorsements". Such a section may not be included in the Modified version.
N. Do not retitle any existing section to be Entitled "Endorsements" or to conflict in title with any Invariant Section.
O. Preserve any Warranty Disclaimers.

If the Modified Version includes new front-matter sections or appendices that qualify as Secondary Sections and contain no material copied from the Document, you may at your option designate some or all of these sections as invariant. To do this, add their titles to the list of Invariant Sections in the Modified Version's license notice. These titles must be distinct from any other section titles.
You may add a section Entitled "Endorsements", provided it contains nothing but endorsements of your Modified Version by various parties—for example, statements of peer review or that the text has been approved by an organization as the authoritative definition of a standard.
You may add a passage of up to five words as a Front-Cover Text, and a passage of up to 25 words as a Back-Cover Text, to the end of the list of Cover Texts in the Modified Version. Only one passage of Front-Cover Text and one of Back-Cover Text may be added by (or through arrangements made by) any one entity. If the Document already includes a cover text for the same cover, previously added by you or by arrangement made by the same entity you are acting on behalf of, you may not add another; but you may replace the old one, on explicit permission from the previous publisher that added the old one.
The author(s) and publisher(s) of the Document do not by this License give permission to use their names for publicity for or to assert or imply endorsement of any Modified Version.

5. COMBINING DOCUMENTS

You may combine the Document with other documents released under this License, under the terms defined in section 4 above for modified versions, provided that you include in the combination all of the Invariant Sections of all of the original documents, unmodified, and list them all as Invariant Sections of your combined work in its license notice, and that you preserve all their Warranty Disclaimers.
The combined work need only contain one copy of this License, and multiple identical Invariant Sections may be replaced with a single copy. If there are multiple Invariant Sections with the same name but different contents, make the title of each such section unique by adding at the end of it, in parentheses, the name of the original author or publisher of that section if known, or else a unique number. Make the same adjustment to the section titles in the list of Invariant Sections in the license notice of the combined work.
In the combination, you must combine any sections Entitled "History" in the various original documents, forming one section Entitled "History"; likewise combine any sections Entitled "Acknowledgements", and any sections Entitled "Dedications". You must delete all sections Entitled "Endorsements".

6. COLLECTIONS OF DOCUMENTS

You may make a collection consisting of the Document and other documents released under this License, and replace the individual copies of this License in the various documents with a single copy that is included in the collection, provided that you follow the rules of this License for verbatim copying of each of the documents in all other respects.
You may extract a single document from such a collection, and distribute it individually under this License, provided you insert a copy of this License into the extracted document, and follow this License in all other respects regarding verbatim copying of that document.

7. AGGREGATION WITH INDEPENDENT WORKS

A compilation of the Document or its derivatives with other separate and independent documents or works, in or on a volume of a storage or distribution medium, is called an "aggregate" if the copyright resulting from the compilation is not used to limit the legal rights of the compilation's users beyond what the individual works permit. When the Document is included in an aggregate, this License does not apply to the other works in the aggregate which are not themselves derivative works of the Document.

If the Cover Text requirement of section 3 is applicable to these copies of the Document, then if the Document is less than one half of the entire aggregate, the Document's Cover Texts may be placed on covers that bracket the Document within the aggregate, or the electronic equivalent of covers if the Document is in electronic form. Otherwise they must appear on printed covers that bracket the whole aggregate.

8. TRANSLATION

Translation is considered a kind of modification, so you may distribute translations of the Document under the terms of section 4. Replacing Invariant Sections with translations requires special permission from their copyright holders, but you may include translations of some or all Invariant Sections in addition to the original versions of these Invariant Sections. You may include a translation of this License, and all the license notices in the Document, and any Warranty Disclaimers, provided that you also include the original English version of this License and the original versions of those notices and disclaimers. In case of a disagreement between the translation and the original version of this License or a notice or disclaimer, the original version will prevail.

If a section in the Document is Entitled "Acknowledgements", "Dedications", or "History", the requirement (section 4) to Preserve its Title (section 1) will typically require changing the actual title.

9. TERMINATION

You may not copy, modify, sublicense, or distribute the Document except as expressly provided under this License. Any attempt otherwise to copy, modify, sublicense, or distribute it is void, and will automatically terminate your rights under this License.

However, if you cease all violation of this License, then your license from a particular copyright holder is reinstated (a) provisionally, unless and until the copyright holder explicitly and finally terminates your license, and (b) permanently, if the copyright holder fails to notify you of the violation by some reasonable means prior to 60 days after the cessation.

Moreover, your license from a particular copyright holder is reinstated permanently if the copyright holder notifies you of the violation by some reasonable means, this is the first time you have received notice of violation of this License (for any work) from that copyright holder, and you cure the violation prior to 30 days after your receipt of the notice.

Termination of your rights under this section does not terminate the licenses of parties who have received copies or rights from you under this License. If your rights have been terminated and not permanently reinstated, receipt of a copy of some or all of the same material does not give you any rights to use it.

10. FUTURE REVISIONS OF THIS LICENSE

The Free Software Foundation may publish new, revised versions of the GNU Free Documentation License from time to time. Such new versions will be similar in spirit to the present version, but may differ in detail to address new problems or concerns. See http://www.gnu.org/copyleft/.

Each version of the License is given a distinguishing version number. If the Document specifies that a particular numbered version of this License "or any later version" applies to it, you have the option of following the terms and conditions either of that specified version or of any later version that has been published (not as a draft) by the Free Software Foundation. If the Document does not specify a version number of this License, you may choose any version ever published (not as a draft) by the Free Software Foundation. If the Document specifies that a proxy can decide which future versions of this License can be used, that proxy's public statement of acceptance of a version permanently authorizes you to choose that version for the Document.

11. RELICENSING

"Massive Multiauthor Collaboration Site" (or "MMC Site") means any World Wide Web server that publishes copyrightable works and also provides prominent facilities for anybody to edit those works. A public wiki that anybody can edit is an example of such a server. A "Massive Multiauthor Collaboration" (or "MMC") contained in the site means any set of copyrightable works thus published on the MMC site.

"CC-BY-SA" means the Creative Commons Attribution-Share Alike 3.0 license published by Creative Commons Corporation, a not-for-profit corporation with a principal place of business in San Francisco, California, as well as future copyleft versions of that license published by that same organization.

"Incorporate" means to publish or republish a Document, in whole or in part, as part of another Document.

An MMC is "eligible for relicensing" if it is licensed under this License, and if all works that were first published under this License somewhere other than this MMC, and subsequently incorporated in whole or in part into the MMC, (1) had no cover texts or invariant sections, and (2) were thus incorporated prior to November 1, 2008.

The operator of an MMC Site may republish an MMC contained in the site under CC-BY-SA on the same site at any time before August 1, 2009, provided the MMC is eligible for relicensing.

How to use this License for your documents

To use this License in a document you have written, include a copy of the License in the document and put the following copyright and license notices just after the title page:

> Copyright (c) YEAR YOUR NAME.
>
> Permission is granted to copy, distribute and/or modify this document
>
> under the terms of the GNU Free Documentation License, Version 1.3
>
> or any later version published by the Free Software Foundation;
>
> with no Invariant Sections, no Front-Cover Texts, and no Back-Cover Texts.
>
> A copy of the license is included in the section entitled "GNU
>
> Free Documentation License".

If you have Invariant Sections, Front-Cover Texts and Back-Cover Texts, replace the "with...Texts." line with this:

> with the Invariant Sections being LIST THEIR TITLES, with the
>
> Front-Cover Texts being LIST, and with the Back-Cover Texts being LIST.

If you have Invariant Sections without Cover Texts, or some other combination of the three, merge those two alternatives to suit the situation.

If your document contains nontrivial examples of program code, we recommend releasing these examples in parallel under your choice of free software license, such as the GNU General Public License, to permit their use in free software.

GNU Free Documentation License Version 1.2, November 2002 Copyright (C) 2000,2001,2002 Free Software Foundation, Inc. 59 Temple Place, Suite 330, Boston, MA 02111-1307 USA Everyone is permitted to copy and distribute verbatim copies of this license document, but changing it is not allowed.

0. PREAMBLE

The purpose of this License is to make a manual, textbook, or other functional and useful document "free" in the sense of freedom: to assure everyone the effective freedom to copy and redistribute it, with or without modifying it, either commercially or noncommercially. Secondarily, this License preserves for the author and publisher a way to get credit for their work, while not being considered responsible for modifications made by others. This License is a kind of "copyleft", which means that derivative works of the document must themselves be free in the same sense. It complements the GNU General Public License, which is a copyleft license designed for free software. We have designed this License in order to use it for manuals for free software, because free software needs free documentation: a free program should come with manuals providing the same freedoms that the software does. But this License is not limited to software manuals; it can be used for any textual work, regardless of subject matter or whether it is published as a printed book. We recommend this License principally for works whose purpose is instruction or reference.

1. APPLICABILITY AND DEFINITIONS

This License applies to any manual or other work, in any medium, that contains a notice placed by the copyright holder saying it can be distributed under the terms of this License. Such a notice grants a world-wide, royalty-free license, unlimited in duration, to use that work under the conditions stated herein. The "Document", below, refers to any such manual or work. Any member of the public is a licensee, and is addressed as "you". You accept the license if you copy, modify or distribute the work in a way requiring permission under copyright law. A "Modified Version" of the Document means any work containing the Document or a portion of it, either copied verbatim, or with modifications and/or translated into another language. A "Secondary Section" is a named appendix or a front-matter section of the Document that deals exclusively with the relationship of the publishers or authors of the Document to the Document's overall subject (or to related matters) and contains nothing that could fall directly within that overall subject. (Thus, if the Document is in part a textbook of mathematics, a Secondary Section may not explain any mathematics.) The relationship could be a matter of historical connection with the subject or with related matters, or of legal, commercial, philosophical, ethical or political position regarding them. The "Invariant Sections" are certain Secondary Sections whose titles are designated, as being those of Invariant Sections, in the notice that says that the Document is released under this License. If a section does not fit the above definition of Secondary then it is not allowed to be designated as Invariant. The Document may contain zero Invariant Sections. If the Document does not identify any Invariant Sections then there are none. The "Cover Texts" are certain short passages of text that are listed, as Front-Cover Texts or Back-Cover Texts, in the notice that says that the Document is released under this License. A Front-Cover Text may be at most 5 words, and a Back-Cover Text may be at most 25 words. A "Transparent" copy of the Document means a machine-readable copy, represented in a format whose specification is available to the general public, that is suitable for revising the document straightforwardly with generic text editors or (for images composed of pixels) generic paint programs or (for drawings) some widely available drawing editor, and that is suitable for input to text formatters or for automatic translation to a variety of formats suitable for input to text formatters. A copy made in an otherwise Transparent file format whose markup, or absence of markup, has been arranged to thwart or discourage subsequent modification by readers is not Transparent. An image format is not Transparent if used for any substantial amount of text. A copy that is not "Transparent" is called "Opaque". Examples of suitable formats for Transparent copies include plain ASCII without markup, Texinfo input format, LaTeX input format, SGML or XML using a publicly available DTD, and standard-conforming simple HTML, PostScript or PDF designed for human modification. Examples of transparent image formats include PNG, XCF and JPG. Opaque formats include proprietary formats that can be read and edited only by proprietary word processors, SGML or XML for which the DTD and/or processing tools are not generally available, and the machine-generated HTML, PostScript or PDF produced by some word processors for output purposes only. The "Title Page" means, for a printed book, the title page itself, plus such following pages as are needed to hold, legibly, the material this License requires to appear in the title page. For works in formats which do not have any title page as such, "Title Page" means the text near the most prominent appearance of the work's title, preceding the beginning of the body of the text. A section "Entitled XYZ" means a named subunit of the Document whose title either is precisely XYZ or contains XYZ in parentheses following text that translates XYZ in another language. (Here XYZ stands for a specific section name mentioned below, such as "Acknowledgements", "Dedications", "Endorsements", or "History".) To "Preserve the Title" of such a section when you modify the Document means that it remains a section "Entitled XYZ" according to this definition. The Document may include Warranty Disclaimers next to the notice which states that this License applies to the Document. These Warranty Disclaimers are considered to be included by reference in this License, but only as regards disclaiming warranties: any other implication that these Warranty Disclaimers may have is void and has no effect on the meaning of this License.

2. VERBATIM COPYING

You may copy and distribute the Document in any medium, either commercially or noncommercially, provided that this License, the copyright notices, and the license notice saying this License applies to the Document are reproduced in all copies, and that you add no other conditions whatsoever to those of this License. You may not use technical measures to obstruct or control the reading or further copying of the copies you make or distribute. However, you may accept compensation in exchange for copies. If you distribute a large enough number of copies you must also follow the conditions in section 3. You may also lend copies, under the same conditions stated above, and you may publicly display copies.

3. COPYING IN QUANTITY

If you publish printed copies (or copies in media that commonly have printed covers) of the Document, numbering more than 100, and the Document's license notice requires Cover Texts, you must enclose the copies in covers that carry, clearly and legibly, all these Cover Texts: Front-Cover Texts on the front cover, and Back-Cover Texts on the back cover. Both covers must also clearly and legibly identify you as the publisher of these copies. The front cover must present the full title with all words of the title equally prominent and visible. You may add other material on the covers in addition. Copying with changes limited to the covers, as long as they preserve the title of the Document and satisfy these conditions, can be treated as verbatim copying in other respects. If the required texts for either cover are too voluminous to fit legibly, you should put the first ones listed (as many as fit reasonably) on the actual cover, and continue the rest onto adjacent pages. If you publish or distribute Opaque copies of the Document numbering more than 100, you must either include a machine-readable Transparent copy along with each Opaque copy, or state in or with each Opaque copy a computer-network location from which the general network-using public has access to download using public-standard network protocols a complete Transparent copy of the Document, free of added material. If you use the latter option, you must take reasonably prudent steps, when you begin distribution of Opaque copies in quantity, to ensure that this Transparent copy will remain thus accessible at the stated location until at least one year after the last time you distribute an Opaque copy (directly or through your agents or retailers) of that edition to the public. It is requested, but not required, that you contact the authors of the Document well before redistributing any large number of copies, to give them a chance to provide you with an updated version of the Document.

4. MODIFICATIONS

You may copy and distribute a Modified Version of the Document under the conditions of sections 2 and 3 above, provided that you release the Modified Version under precisely this License, with the Modified Version filling the role of the Document, thus licensing distribution and modification of the Modified Version to whoever possesses a copy of it. In addition, you must do these things in the Modified Version: A. Use in the Title Page (and on the covers, if any) a title distinct from that of the Document, and from those of previous versions (which should, if there were any, be listed in the History section of the Document). You may use the same title as a previous version if the original publisher of that version gives permission. B. List on the Title Page, as authors, one or more persons or entities responsible for authorship of the modifications in the Modified Version, together with at least five of the principal authors of the Document (all of its principal authors, if it has fewer than five), unless they release you from this requirement. C. State on the Title page the name of the publisher of the Modified Version, as the publisher. D. Preserve all the copyright notices of the Document. E. Add an appropriate copyright notice for your modifications adjacent to the other copyright notices. F. Include, immediately after the copyright notices, a license notice giving the public permission to use the Modified Version under the terms of this License, in the form shown in the Addendum below. G. Preserve in that license notice the full lists of Invariant Sections and required Cover Texts given in the Document's license notice. H. Include an unaltered copy of this License. I. Preserve the section Entitled "History", Preserve its Title, and add to it an item stating at least the title, year, new authors, and publisher of the Modified Version as given on the Title Page. If there is no section Entitled "History" in the Document, create one stating the title, year, authors, and publisher of the Document as given on its Title Page, then add an item describing the Modified Version as stated in the previous sentence. J. Preserve the network location, if any, given in the Document for public access to a Transparent copy of the Document, and likewise the network locations given in the Document for previous versions it was based on. These may be placed in the "History" section. You may omit a network location for a work that was published at least four years before the Document itself, or if the original publisher of the version it refers to gives permission. K. For any section Entitled "Acknowledgements" or "Dedications", Preserve the Title of the section, and preserve in the section all the substance and tone of each of the contributor acknowledgements and/or dedications given therein. L. Preserve all the Invariant Sections of the Document, unaltered in their text and in their titles. Section numbers or the equivalent are not considered part of the section titles. M. Delete any section Entitled "Endorsements". Such a section may not be included in the Modified Version. N. Do not retitle any existing section to be Entitled "Endorsements" or to conflict in title with any Invariant Section. O. Preserve any Warranty Disclaimers. If the Modified Version includes new front-matter sections or appendices that qualify as Secondary Sections and contain no material copied from the Document, you may at your option designate some or all of these sections as invariant. To do this, add their titles to the list of Invariant Sections in the Modified Version's license notice. These titles must be distinct from any other section titles. You may add a section Entitled "Endorsements", provided it contains nothing but endorsements of your Modified Version by various parties--for example, statements of peer review or that the text has been approved by an organization as the authoritative definition of a standard. You may add a passage of up to five words as a Front-Cover Text, and a passage of up to 25 words as a Back-Cover Text, to the end of the list of Cover Texts in the Modified Version. Only one passage of Front-Cover Text and one of Back-Cover Text may be added by (or through arrangements made by) any one entity. If the Document already includes a cover text for the same cover, previously added by you or by arrangement made by the same entity you are acting on behalf of, you may not add another; but you may replace the old one, on explicit permission from the previous publisher that added the old one. The author(s) and publisher(s) of the Document do not by this License give permission to use their names for publicity for or to assert or imply endorsement of any Modified Version.

5. COMBINING DOCUMENTS

You may combine the Document with other documents released under this License, under the terms defined in section 4 above for modified versions, provided that you include in the combination all of the Invariant Sections of all of the original documents, unmodified, and list them all as Invariant Sections of your combined work in its license notice, and that you preserve all their Warranty Disclaimers. The combined work need only contain one copy of this License, and multiple identical Invariant Sections may be replaced with a single copy. If there are multiple Invariant Sections with the same name but different contents, make the title of each such section unique by adding at the end of it, in parentheses, the name of the original author or publisher of that section if known, or else a unique number. Make the same adjustment to the section titles in the list of Invariant Sections in the license notice of the combined work. In the combination, you must combine any sections Entitled "History" in the various original documents, forming one section Entitled "History"; likewise combine any sections Entitled "Acknowledgements", and any sections Entitled "Dedications". You must delete all sections Entitled "Endorsements".

6. COLLECTIONS OF DOCUMENTS

You may make a collection consisting of the Document and other documents released under this License, and replace the individual copies of this License in the various documents with a single copy that is included in the collection, provided that you follow the rules of this License for verbatim copying of each of the documents in all other respects. You may extract a single document from such a collection, and distribute it individually under this License, provided you insert a copy of this License into the extracted document, and follow this License in all other respects regarding verbatim copying of that document.

7. AGGREGATION WITH INDEPENDENT WORKS

A compilation of the Document or its derivatives with other separate and independent documents or works, in or on a volume of a storage or distribution medium, is called an "aggregate" if the copyright resulting from the compilation is not used to limit the legal rights of the compilation's users beyond what the individual works permit. When the Document is included in an aggregate, this License does not apply to the other works in the aggregate which are not themselves derivative works of the Document. If the Cover Text requirement of section 3 is applicable to these copies of the Document, then if the Document is less than one half of the entire aggregate, the Document's Cover Texts may be placed on covers that bracket the Document within the aggregate, or the electronic equivalent of covers if the Document is in electronic form. Otherwise they must appear on printed covers that bracket the whole aggregate.

8. TRANSLATION

Translation is considered a kind of modification, so you may distribute translations of the Document under the terms of section 4. Replacing Invariant Sections with translations requires special permission from their copyright holders, but you may include translations of some or all Invariant Sections in addition to the original versions of these Invariant Sections. You may include a translation of this License, and all the license notices in the Document, and any Warranty Disclaimers, provided that you also include the original English version of this License and the original versions of those notices and disclaimers. In case of a disagreement between the translation and the original version of this License or a notice or disclaimer, the original version will prevail. If a section in the Document is Entitled "Acknowledgements", "Dedications", or "History", the requirement (section 4) to Preserve its Title (section 1) will typically require changing the actual title.

9. TERMINATION

You may not copy, modify, sublicense, or distribute the Document except as expressly provided for under this License. Any other attempt to copy, modify, sublicense or distribute the Document is void, and will automatically terminate your rights under this License. However, parties who have received copies, or rights, from you under this License will not have their licenses terminated so long as such parties remain in full compliance.

10. FUTURE REVISIONS OF THIS LICENSE

The Free Software Foundation may publish new, revised versions of the GNU Free Documentation License from time to time. Such new versions will be similar in spirit to the present version, but may differ in detail to address new problems or concerns. See http://www.gnu.org/copyleft/. Each version of the License is given a distinguishing version number. If the Document specifies that a particular numbered version of this License "or any later version" applies to it, you have the option of following the terms and conditions either of that specified version or of any later version that has been published (not as a draft) by the Free Software Foundation. If the Document does not specify a version number of this License, you may choose any version ever published (not as a draft) by the Free Software Foundation. ADDENDUM: How to use this License for your documents To use this License in a document you have written, include a copy of the License in the document and put the following copyright and license notices just after the title page: Copyright (c) YEAR YOUR NAME. Permission is granted to copy, distribute and/or modify this document under the terms of the GNU Free Documentation License, Version 1.2 or any later version published by the Free Software Foundation; with no Invariant Sections, no Front-Cover Texts, and no Back-Cover Texts. A copy of the license is included in the section entitled "GNU Free Documentation License". If you have Invariant Sections, Front-Cover Texts and Back-Cover Texts, replace the "with...Texts." line with this: with the Invariant Sections being LIST THEIR TITLES, with the Front-Cover Texts being LIST, and with the Back-Cover Texts being LIST. If you have Invariant Sections without Cover Texts, or some other combination of the three, merge those two alternatives to suit the situation. If your document contains nontrivial examples of program code, we recommend releasing these examples in parallel under your choice of free software license, such as the GNU General Public License, to permit their use in free software.